MISSION TO
THE PLANETS

FRONTISPIECE

Neptune and Triton; shown together two days,
six and a half hours after closest approach,
from 3,020,000 miles (4,860,000 km).
Voyager is now plunging southward at
48°S to the plane of the ecliptic.

MISSION TO THE PLANETS

The Illustrated Story of Man's
Exploration of the Solar System

PATRICK MOORE

CASSELL

Also by Patrick Moore:

The Planet Neptune

The Amateur Astronomer

Astronomers' Stars

Stargazing

Exploring the Night Sky with Binoculars

The Guinness Book of Astronomy

Atlas of the Universe

A CASSELL BOOK

Cassell Publishers Limited
Wellington House, 125 Strand, London WC2R 0BB

Copyright © Patrick Moore 1990, 1995

First published 1990
First paperback edition 1991
Second edition 1995

Distributed in the United States
by Sterling Publishing Co., Inc.
387 Park Avenue South,
New York, New York 10016-8810

Distributed in Australia
by Capricorn Link (Australia) Pty Ltd
2/13 Carrington Road, Castle Hill, NSW 2154

British Library Cataloguing in Publication Data

A catalogue record for this book is available from the British Library

ISBN 0-304-34603-9

Typeset by Fakenham Photosetting Ltd
Fakenham, Norfolk
Printed and bound in Singapore

CONTENTS

PREFACE

The Space Age began little more than thirty years ago. Since then so much has happened that in this book I cannot hope to do more than give a brief summary. At least my memory goes back a long way, to a time well before space-flight was taken seriously except by a few enthusiasts; I have mapped the Moon, I have been present at most of the planetary encounters, and I know most of the people involved. I can only hope that what I have to say will be of some interest.

Since both Imperial and Metric are in common use, I have thought it best to give both, but in some cases I have rounded off to the nearest convenient whole numbers.

All the photographs in this book are reproduced by courtesy of NASA, apart from the photographs from Russian space-craft, which are reproduced by courtesy of the USSR Academy of Sciences.

My thanks are due to Paul Doherty, for his splendid illustrations; to Rosemary Anderson for her invaluable guidance in seeing the book through the Press, to Simon Bell for his excellent design and to Matthew Clarke, who was initially responsible for this book.

Patrick Moore
Selsey

PREFACE TO THE SECOND EDITION

Much has happened during the five years since the first edition of this book was written, and I have therefore taken the opportunity to bring the text fully up to date.

Patrick Moore
Selsey

1

INTRODUCTION: TO OTHER WORLDS

On the evening of 21 July, 1969, I was sitting in a television studio in London. This was Lime Grove – not the BBC Television Centre – and the monitor in front of me was black and white. I was there for a special reason. My task was to give a commentary as Neil Armstrong and Edwin Aldrin, both of whom I knew, made their way down on to the surface of the Moon. Their voices came through as clearly as if they had been in the next studio.

All sorts of thoughts flashed through my mind. As one of the 'Moon mappers' who had spent many years in charting the lunar world, I knew what lay below the descending space-craft; but after all, what had we really found out about the Moon itself? There had been a popular theory that the so-called 'seas' were deep dust-drifts, and if the space-craft came down in a treacherous area, or landed at a steep angle, the result could only be disaster. There was no provision for rescue, and nobody could survive for long upon an airless, lifeless world.

The countdown went on . . . and then, suddenly, I heard Neil's voice: 'The *Eagle* has landed.' The feeling of relief was tremendous. I am not sure what I said – I was 'on the air', so I hope that it made sense; in any case, I am quite sure that the same sense of relief was felt throughout the world. I had been privileged to be present at the moment of one of the greatest triumphs in the history of mankind.

The idea of space-flight was not new. It went back to Ancient Greece, and various weird and wonderful methods of reaching the Moon had been proposed, ranging from waterspouts to demons, space-guns and gigantic catapults. But only in our own century had it become possible to harness the power of the rocket, which alone can travel in empty space above the top of the Earth's atmosphere. The Space Age itself was a mere dozen years old, even though its opening, with the flight of Russia's artificial satellite Sputnik 1 on 4 October 1957, already seemed a long time ago.

There had always been intense interest in our neighbour worlds, and over the centuries we had found out most of the basic facts. We had learned that the Sun is a star; that many of the other stars you can see on any clear night are a great deal larger, hotter and more luminous than our Sun, and that there are around a hundred thousand million stars in the Galaxy, beyond which we can make out other galaxies so remote that their light takes millions, hundreds of millions or even thousands of millions of years to reach us. We had found, too, that the Sun's family or Solar System contains nine planets, of which the Earth is one; some are larger than the Earth, while others are smaller. The Moon, which looks so splendid in our sky, is a very insignificant body, staying together with us as we travel round the Sun and shining by reflecting the solar rays in the manner of a rather inefficient mirror. But up to now we had had to depend upon what we could discover by looking at other worlds across space; even the Moon is around 239,000 miles (384,000 km) away. The flight of Apollo 11 meant that man had at last found how to travel from his own world to another.

To me, the touch-down on the Moon was the most significant moment of all. It even dwarfed the experience of a few hours later, when first Neil Armstrong, then Edwin Aldrin, clambered down the ladder at the side of the grounded *Eagle*, and we heard the never-to-be-forgotten words: 'That's one small step for a man, one giant leap for mankind.' There was another moment of tension when the time came for blasting back into orbit. The lunar module had only one ascent engine, and this engine had to work faultlessly, first time; there could be no second chance.

It had all happened much more quickly than most people had expected. The first modern-type rocket had been fired in 1926, by Robert Hutchings Goddard in America, and though it was a tiny thing, which moved at a top speed of less than 60 mph (100 kph) and covered no more than 200 ft (60 m) before crash-landing, it was the true ancestor of the space-craft which had sent the Apollo astronauts on their way to the Moon. In 1957 came the first satellite, Sputnik 1, followed four years later by the flight of Yuri Gagarin in his cramped Vostok module. Since then there had been elaborate space vehicles which were able to send back amazing views of the Earth itself and also to carry equipment which could tell us more about the regions of near space. We had been able to study radiations which are inaccessible from ground level because they are blocked out by layers in the upper atmosphere; we had discovered the zones of radiation round the Earth which are today known as the Van Allen zones; we had collected information from remote stars and star-systems, and we had also put the satellites to practical use. In 1950, for example, who could have predicted that within a couple of decades it would be possible to switch on a television set in London and watch a game of baseball being played in New York?

Planets were not neglected, and even before the Apollo mission there had been remarkable successes. Venus was first on the list. It had been thought to be a warm, perhaps welcoming world, with oceans which might well support life of the type which flourished on Earth during the Coal-Forest period, but Mariner 2, which flew past Venus in 1962, told a different story. Next there was Mars, the Red Planet; did it support life – and what was the truth about the strange, artificial-looking streaks which had become known as canals, and which Percival Lowell and others had believed to be genuine waterways, built by brilliant engineers to form a vast irrigation system? Even if the canal-builders were discounted, there still seemed every chance that life of a sort could survive on Mars, though we could not decide one way or the other before the first fly-by mission, Mariner 4 of 1965. Further from the Sun there were the giant planets, Jupiter and the rest, with their satellite families and their deep, gaseous

Approaching the Moon. This photograph was taken from *Eagle*, the landing craft of Apollo 11, just before touch-down. The shadow of the space-craft is seen on the lunar surface. The well-defined crater is named Maskelyne. Photograph by Neil Armstrong.

atmospheres. Neither could we forget Halley's Comet, which was already on its way back to the Sun and would be within range in 1986. What was it really like, and would there be any chance of sending a space probe to find out?

This, remember, was less than thirty years ago. By now we have sent seven expeditions to the Moon, of which only one has failed, and space-craft have flown past all the planets apart from Pluto, while controlled landings have been made on Mars and Venus. We have found no life, but there has been plenty to intrigue us, and it is fair to say that each planet has produced its quota of surprises.

What I hope to do in this book is to tell the story of the first decades of planetary exploration. It is bound to be a somewhat personal account – but after all I have lived through it, and I have known most of the people concerned. First, however, we must take a general look at the Solar System, our home in space.

2

OUR PART OF THE UNIVERSE

The Solar System is dominated by the Sun, which may be nothing more than a normal star, but which is much more massive than all the planets put together. It is a globe of incandescent gas, large enough to swallow up more than a million bodies the volume of the Earth, and it is extremely hot. Even its surface is at a temperature not far short of 11,000°F (6,000°C), and near its core the temperature rises to the almost incredible value of at least 25,000,000°F (14,000,000°C).

Without the Sun, there would be no Solar System. It seems almost certain that the planets were formed from what is termed a solar nebula – that is to say, a cloud of material associated with the youthful Sun, out of which the planets built up by the process of accretion. We know the age of the Earth with fair certainty: it is about 4,600 million years, and the Sun is presumably older, so that if we date the Sun back to 5,000 million years we are probably not far wrong.

The Sun is not burning in the manner of a coal fire. It is producing its energy by nuclear reactions going on deep inside it, so that, in a way, it may be likened to a huge, controlled nuclear bomb. One gas (hydrogen) is being changed into another (helium), so that hydrogen is the Sun's fuel. It takes four nuclei of hydrogen to form one nucleus of helium, and each time this happens a little energy is set free, which is why the Sun shines. The process is a rather roundabout one, and it also involves loss of mass, so that the Sun is losing 4,000,000 tons every second; it 'weighs' much less now than it did when you started reading this page. I hasten to add that there is no immediate cause for alarm. The Sun contains so much material that it will not change much for several thousands of millions of years in the future, and by stellar standards it is no more than middle-aged.

(It is rather interesting to recall that Sir William Herschel, arguably the greatest observational astronomer in history, thought that below the Sun's surface there was a cool, pleasant region which might well be inhabited. He continued to believe this until the end of his life in 1822. Not many of his contemporaries agreed with him, and I would certainly not recommend trying to go there to find out.)

As the Sun shines, it sends out streams of atomic particles which make up what is termed the solar wind. It has been known for some time, though its existence was not finally proved until results came back from interplanetary space-craft. It has a pronounced effect upon the tails of comets, and it can also overload the Van Allen radiation zones surrounding the Earth, so that particles cascade down into the upper air and produce the lovely sky-glows which are called auroræ or polar lights. Because the particles are electrified, they tend to cascade down towards the magnetic poles, which is why the best auroral displays are seen from high latitudes. They are associated, too, with solar flares, which are violent, short-lived outbreaks on the Sun and make the solar wind 'gusty'. Flares often occur over sunspots, which are cooler areas on the solar surface and are the centres of powerful magnetic fields. They are not permanent, for obvious reasons; the Sun is to some extent a variable star, and every eleven years or so it is particularly energetic, so that spots (and auroræ) are common. The last maximum fell in 1990, and although we cannot predict the next maximum accurately it should be timed for some time around 2001.

Auroræ are familiar features of the night sky from places such as Scotland or Norway. From southern England they are less frequent, but there was a particularly brilliant display on 13 March 1989, when the whole sky glowed vividly. I saw it from my home in Sussex; at the peak of the display I had a telephone call from Paul Doherty, a colleague living in Staffordshire, telling me that from there the aurora was bright enough to cast shadows, with streamers, arcs, curtains and rapidly changing hues. It was not entirely unexpected, because at that time there was a huge sunspot group on view. Of course, auroræ occur in the far south as well as in the north; this particular display was well seen from parts of New Zealand.

The planets move round the Sun at various distances, in periods ranging from 88 Earth-days for Mercury up to 248 years for Pluto. The traffic laws of the Solar System state that the closer-in a planet is, the faster it moves; Mercury flashes along at a mean velocity of 30 miles (48 km) per second, while the Earth moves at the more sedate pace of just under 19 miles (30 km) per second on average, while Pluto crawls at a mere 3 miles (4.7 km) per second.

Any casual glance at a plan of the Solar System shows that it is divided into two definite parts. First we have four relatively small planets: Mercury, Venus, the Earth and Mars. Then comes a wide gap, in which move

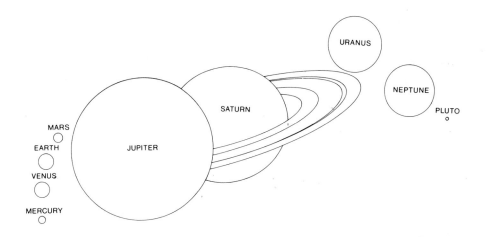

Comparative sizes of the planets. The difference in scale between the four giants and the 'terrestrial' planets is very striking.

thousands of dwarf worlds, known as asteroids or minor planets. Further out we come to the giants, Jupiter, Saturn, Uranus and Neptune, which are totally unlike the members of the inner group; instead of being solid, they have gaseous surfaces and only comparatively small cores. Jupiter, the most important of them, is more massive than all the other planets combined.

Finally we meet Pluto, which seems to be in a class of its own. It is very small – not only smaller than the Earth, but even smaller than the Moon – and it has a strange orbit, much less circular than those of the other planets, so that when near perihelion (that is to say, its closest point to the Sun) it is actually closer-in than Neptune. The last perihelion took place in 1989, and so for the moment Neptune, not Pluto, is the outermost planet; it will remain so until 1999. However, there is no fear of a collision. Most of the planetary orbits lie much in the same plane, so that if you draw a map of the Solar System on a flat piece of paper, you are not very far wrong, but Pluto is again an exception; its orbital tilt is 17 degrees, so that at the present epoch it cannot go anywhere near Neptune. Altogether, Pluto is a maverick, and there are grounds for doubting whether it is worthy of true planetary status.

Most of the planets are attended by satellites. The Earth, of course, has one: our familiar Moon, which seems admittedly rather too large to be classed as a mere satellite, and might be better termed a companion planet. Represent the Earth by a tennis ball, and the Moon will be the size of a table-tennis ball. Of the rest, Saturn has 18 known satellites, Jupiter 16, Uranus 15, Neptune eight and Mars two. Pluto is unusual in this respect also; its companion, Charon, has more than half Pluto's diameter, and the two make up a curious sort of pair.

The first five planets have been known since very ancient times. This is not surprising, because even though they have no light of their own, and depend upon reflecting the rays of the Sun, they are bright naked-eye objects. Pride of place must go to Venus, which can even cast shadows at times; it is understandable that it was named in honour of the Goddess of Beauty. Jupiter also is much brighter than any star; Mars can become brilliant, and is easy to recognize because of its red colour, and even Saturn is bright enough to be conspicuous. Mercury is less in evidence, because it can never be seen against a dark sky, but when at its best it is not hard to find.

Seven was the mystical number of the ancients, and so it was natural for there to be seven bodies in the Solar System: the five known planets, plus the Sun and the Moon. It came as a surprise when a new planet was discovered. It was found in 1781 by William Herschel, a Hanoverian-born musician who had settled in England and had made his own telescopes; after some discussion it was named Uranus, after the first ruler of Olympus. It is just visible with the naked eye if you know where to look for it, but until the Space Age we knew remarkably little about it.

Aurora Borealis, as seen from Scotland. The colours were vivid, as is often the case with auroræ.

When a new body is discovered, the mathematicians set to work and decide how it ought to move. Uranus proved to be un-cooperative. It persistently wandered away from its predicted path, and the cause of the trouble was tracked down to the pull of a more distant planet, which was located in 1846 just where the mathematicians had said it should be; following the mythological pattern, it was named Neptune, in honour of the sea-god. It is much too faint to be seen with the naked eye, but binoculars will show it easily. Calculations of the same sort led to the discovery of Pluto, by my old friend Clyde Tombaugh, in 1930. Whether or not the discovery had an element of luck about it remains to be decided; in any case, Pluto cannot be seen without a fairly powerful telescope.

The asteroids, most of which keep strictly to the region of the Solar System between the orbits of Mars and Jupiter, are very small. Only one (Ceres) is as much as 580 miles (940 km) in diameter, and only one (Vesta) is ever visible with the naked eye. It used to be thought that they represented the débris of an old planet which met with disaster in the remote past, but it now seems more likely that no planet could ever form in this part of the Solar System because of the powerful disruptive pull of Jupiter; as soon as a large planet started to build up, Jupiter tore it to pieces again, so that the end-product was a swarm of dwarfs.

Though the larger asteroids keep strictly to the main belt beyond the path of Mars, others do not, and may come close to the Earth. In 1937, for example, the asteroid Hermes, half a mile (1 km) wide, passed at only twice the distance of the Moon. This record has since been broken many times, and we have found tiny asteroids which have even passed between the Earth and the Moon. Inevitably it has been suggested that one of these midgets could hit us, and the chances of collision cannot be ruled out, particularly as the so-called 'near-Earth objects' are much commoner than we used to think, and new discoveries are being made every year. It has even been claimed that an asteroid hit, 65,000,000 years ago, changed the climate to such an extent that the dinosaurs could not cope with the new conditions, and died out. This may or may not be true (personally I am somewhat sceptical), but it is true to say that we are not immune.

Apart from the planets, satellites and asteroids, the Solar System contains bodies of different type. In particular, there are the comets, which may look spectacular, but are very flimsy; I have described a comet as being 'the nearest approach to nothing which can still be anything'. The only substantial part is the central nucleus, which is made up chiefly of ice, and is only a few miles across.

Most comets travel in very long, narrow paths, so that they spend most of their time a long way from the Sun; they are then too dim to be seen, and are nothing more than icy lumps. When the comet moves inwards, and is heated, the ices in the nucleus start to evaporate, so that the comet develops a head or coma. Tails are of two types, dusty and gaseous. Dust-tails are usually curved, while gas-tails are straight. Both point more or less away from the Sun, because they are pushed outward, by radiation pressure in the case of dust-tails and by the solar wind in the case of gas-tails. When a comet is moving away from the Sun, it travels tail-first.

Bright comets have very long periods, amounting to many centuries, so

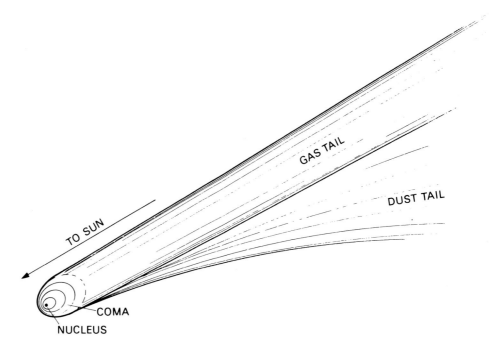

Anatomy of a comet. The icy nucleus is the only substantial part; around it is the coma, and some comets have both gas or ion tails (straight) and dust tails (curved).

that we never know when to expect them, and they are always apt to take us by surprise. The only exception is Halley's Comet, which comes back every 76 years, and last returned to perihelion in 1986; it will be back once more in the year 2061. All comets with shorter periods are faint, and few of them develop tails. By cosmical standards comets are short-lived; some have been seen to break up and vanish, and others have hit the Sun, while in 1994 a faint comet, Shoemaker-Levy 9, committed suicide by crashing into Jupiter.

As a comet moves, it leaves a trail of débris behind it. When the Earth passes through one of these trails, it collects numbers of particles, which dash into the upper atmosphere and burn away by friction against the air-particles, producing the luminous streaks which we call shooting-stars or meteors. Meteors are tiny, usually smaller than grains of sand, and burn away well before they can reach ground level. There are various yearly showers, of which the best is that of early August; look up into a dark, clear sky for a few minutes any time between about 27 July and 17 August, and you will be unlucky not to see at least a couple of meteors.

Larger objects can survive the complete drop to the ground, and are then known as meteorites. Most museums have meteorite collections, and we know a great deal about them; some are stony, while others are made principally of iron. Meteorites, please note, are quite different from shooting-star meteors, because they are not associated with comets, and probably come from the asteroid belt, so that there is no real difference between a large meteorite and a small asteroid. Craters may be formed; go to Arizona, not far from the town of Winslow, and you will see a large crater which was produced by the impact of a meteorite which landed there well over 20,000 years ago.

Such is the Solar System. Now let us look back to the start of the Space Age, and see what we then knew – or, rather, what we thought we knew!

3

THE SUN'S FAMILY:
OUR VIEWS IN 1957

In the spring of 1957 I gave a lecture to an audience at Cambridge University. The subject was 'The Planets As We Know Them'. I made a series of 12 positive statements, each of which was backed up by the best available evidence – and nine of which turned out to be wrong. This was not really my fault, particularly as I had been busily engaged in observing the planets for at least a quarter of a century; in many cases the planets were not the sorts of worlds we had believed them to be, and without using space-probes there was no way of finding out.

Of course, some facts were well established. We knew the movements of the planets very precisely; we had mapped the surfaces of those which showed permanent or at least semi-permanent features; in most cases we knew the rotation periods, though our estimates for Mercury and Venus later proved to be wildly in error. But our information was limited so long as we could do no more than look at the planets from distances of many millions of miles.

The Moon is a special case, because it is so close to us. Fly ten times round the Earth's equator, and you will cover a distance which is greater than that between the Earth and the Moon. Also, the lunar surface is always sharp and clear-cut, because there is no atmosphere to cause clouds or mist. The reason for this is not in the least mysterious. The Earth's escape velocity – that is to say, the speed which an object must be given if it is to break free without further impetus – is 7 miles (11 km) per second. Air is made up of millions upon millions of tiny particles, all flying about at high speeds, but they cannot work up to speeds of 7 miles per second, so that the Earth – luckily for us – is able to hold them down. Not so with the Moon, where the escape velocity is a mere 1.5 miles (2.4 km) per second; an Earth-type atmosphere would long since have leaked away into space.

However, in 1957 it was thought that there might be a trace of atmosphere left. The ground density could not be more than around 1/10,000 that of the Earth's air at sea-level, but this would be enough to burn up meteors and produce shooting-star effects; one famous astronomer, Ernst J. Öpik of Armagh, wrote to me in 1956 saying that in his view lunar meteors were 'very probable'. An atmosphere of this density might also have been enough to filter out some of the undesirable short-wave radiations coming from space.

There was also the question of the origin of the walled formations which cover so much of the Moon. Some authorities believed them to be due to the impacts of meteorites, while others regarded them as volcanic. On the whole, the bombardment theory was the more popular, particularly in the United States; 'vulcanists', such as myself, were (and still are) in the minority, though by now it has become clear that both theories have a good deal of truth in them.

At least we were sure that almost nothing had happened on the Moon for a very long time. Even the most recent of the large craters must have been formed well before any advanced life appeared on the Earth, and date back at least a thousand million years. However, there was another theory which was taken very seriously indeed, and which would, if valid, have made lunar landings almost impossible. According to Dr Thomas Gold and his colleagues at Cambridge University, the lunar 'seas' or maria were filled with soft dust. Gold even claimed that a space-ship incautious enough to land there would simply sink out of sight.

I had no faith in this idea, because it did not seem to fit the facts, but there was always a nagging doubt, and obviously no manned expeditions could be sent until the 'Gold-dust' theory had been tested. In 1957, we could only wait and see.

Another problem concerned the far side of the Moon. It is not really correct to say simply that the Moon moves round the Earth; more accurately, the Earth and the Moon move together round their common centre of gravity, which is known as the barycentre. True, the barycentre lies well inside the Earth's globe, because the Earth is 81 times as massive as the Moon, but the distinction is important to mathematicians, and may provide extra support for the idea that the Moon should be ranked as a secondary planet rather than as a satellite.

The Moon's orbital period is 27.3 days, and it spins on its axis in exactly

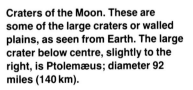

Craters of the Moon. These are some of the large craters or walled plains, as seen from Earth. The large crater below centre, slightly to the right, is Ptolemæus; diameter 92 miles (140 km).

the same time: 27.3 days. This is no mere coincidence, and we know that tidal forces over the ages have been responsible – all other large planetary satellites have similarly synchronous or 'captured' rotations relative to their parent planets – but it means that the same hemisphere of the Moon always faces us, and there is a large area which we can never see from Earth. It was infuriating for lunar observers, but before the space-probe era there was nothing we could do about it.

Actually, we can examine more than half the total surface. Though the Moon spins on its axis at a constant rate, its velocity in orbit changes, because it moves quickest when it is closest to us. This means that there is a slight, slow 'tipping', and all in all almost 60 per cent of the surface can be seen at one time or another; but the 'edges' of the disk are very foreshortened and difficult to map. It is often impossible to tell the difference between a crater wall and a mountain ridge, for example. My own work had been concentrated on trying to chart these foreshortened regions, but I did not claim that the results were particularly good.

There were all sorts of strange speculations about the Moon's hidden regions. (George Adamski, co-author of the first famous book about flying saucers, told me in all seriousness that he had been on a round trip and seen little furry animals running about in a green, pleasant landscape.) However, it seemed reasonable to assume that the far side was just as bleak and lifeless as the side we had always known.

The planets are much further away than the Moon; Mars, for instance, can never approach us much within 36,000,000 miles (59,000,000 km). I recall an apt comment by Gerard de Vaucouleurs, one of the world's leading planetary astronomers, to the effect that no telescope ever built up to that time could show Mars more clearly than the Moon as seen through low-power binoculars. However, definite surface details could be seen. Mars has a generally red surface, with dark patches here and there, and white caps covering the poles; the polar caps wax and wane with the seasons, so that they are largest in Martian summer and smallest in Martian winter. The seasons there are of the same type as ours, apart from being much longer. Mars has a 'year' of 687 Earth-days; this is equivalent to 668 Martian days or 'sols', because Mars spins rather more slowly than the Earth, and has a rotation period of 24 hours 37½ minutes.

There was no real difficulty in charting the main dark patches. Maps were drawn up, and the features were named. Early systems of nomenclature were nothing if not bizarre; for example, the main dark area was called the Hourglass Sea, while a broad reddish region was named Beer Continent (in honour of the German astronomer Wilhelm Beer, who had compiled one of the first reasonably good charts). The modern names are more sober; the Hourglass Sea has become Syrtis Major, while Beer Continent is now known as Aeria.

But could there be lands and seas on Mars?

The red areas were usually called deserts, but it was thought that instead of being sandy they were coated with reddish, rusty minerals. The dark areas might look like seas, but the existence of large oceans was ruled out by the thinness and dryness of the atmosphere, and it seemed more likely that we were dealing with old sea-beds filled with primitive vegetation. It was

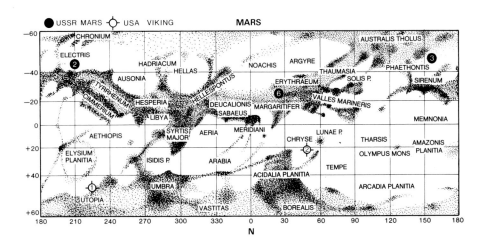

Map of Mars, with impact and landing points of Russian Mars probes and American Vikings 1 and 2. The map is based on the International Astronomical Union map of Mars.

argued that if the patches were not made up of something which could push wind-blown dust aside, they would soon be covered up – and Martian dust-storms are common; I have seen plenty of them over the years, and there are times when the entire planet is covered by a dusty veil.

All this sounded very convincing, and up to 1964, when the first space-craft passed by the planet, most people thought that life of some kind or other must survive there. On the other hand, there was always the problem of the tenuous atmosphere. It was assumed to be made up chiefly of nitrogen, which of course makes up 78 per cent of the air that you and I are breathing, but it was not expected that there would be much free oxygen. The estimated ground pressure was 85 millibars, which is equivalent to the pressure in the Earth's air at a height rather less than twice that of Everest. In other words, human beings or Earth-type animals could not live there, but some types of reasonably advanced life-forms could probably manage.

The polar caps were not thought to be the same as ours, and were dismissed as very thin layers of hoar-frost. There was even a theory that they were made up of 'dry ice' (solid carbon dioxide) rather than water ice, in which case Mars would have been a very dry world indeed. There was no evidence of mountains, craters or valleys; the landscape was taken to be more or less level, or at most gently undulating.

Next, what about the famous – or infamous – canals? Interest in them dated back to 1877, when the Italian observer Giovanni Schiaparelli had drawn them as regular, artificial-looking lines crossing the deserts. He had called them 'canali', or channels, and although he was cautious about their nature he kept an open mind. Percival Lowell, founder of the great observatory at Flagstaff in Arizona, was much more definite. To him, Mars was the home of a brilliant race of beings who were doing their best to survive upon a world which was desperately short of water. Ice from the poles was melted, and pumped through to the warmer regions where the 'Martians' lived, so that a canal was made up of a narrow watercourse surrounded on either side by irrigated land. Lowell's drawings showed a spider's web network of canals, crossing not only the deserts but also the dark regions.

If Lowell's drawings had been accurate, then Mars would have been inhabited; a network of this kind could not possibly be natural. But other observers either could not see the canals at all, or else drew them as vague, ill-defined streaks. There was much better agreement about a seasonal 'wave of darkening'; when a polar cap shrank in the Martian spring, the adjacent dark areas became sharper and more prominent, as though vegetation were being revived by moisture wafted down by the winds.

The main telescope at the Lowell Observatory was a 24-inch (61-cm) refractor, unquestionably one of the best in the world. Perhaps I may be allowed to make a few personal comments here, because I know that telescope well, and have carried out a good deal of work with it. I have never been able to see canals on Mars, and neither have I been able to track the seasonal wave of darkening; I have been equally unsuccessful with other telescopes, ranging from giant instruments down to the modest equipment in my own observatory. Even in Lowell's day, the canal network was regarded with considerable scepticism, but doubts remained, and as recently as 1956 Gérard de Vaucouleurs was still maintaining that the canals were not pure illusions; they had at least a basis of reality.

Two satellites of Mars, Phobos and Deimos, had been discovered in 1877. Both are less than 20 miles (30 km) across, and may well be captured asteroids rather than true satellites, but in some ways they are unusual; Phobos, the inner, has a revolution period of only 7½ hours, so that it completes three orbits every Martian day, and an observer on the surface of Mars would see it rise in the west and set towards the east. At about the time of Sputnik 1, a famous Russian astronomer, Iosif Shklovsky, made the curious suggestion that Phobos and Deimos might be space-stations, launched by the Martians for reasons of their own. The Soviet Academy of Sciences was not impressed with this idea; years later Shklovsky told me that it had never been more than a practical joke, but it was not classed as such at the time!

Venus can come much closer to us than Mars can ever do, but it is much less easy to study, because when at its nearest it is more or less between the Earth and the Sun, so that its dark side is turned towards us and we cannot see it at all – unless the alignment is perfect, when Venus passes in transit across the Sun's face, as last happened in 1882 and will happen again in the year 2004. When full, Venus is on the far side of the Sun. Even when the planet is best placed, the disk usually appears more or less blank, because of the dense, cloud-laden atmosphere which never clears sufficiently to give us a glimpse of the surface. I have made many hundreds of drawings of Venus, but all I can make out are vague, elusive features which shift and change. There is no hope of trying to derive a rotation period, as can be done with Mars. Of course, many observers have tried; some years ago I listed 103 estimates that had been made up to 1955, and not one turned out to be correct!

Analysis of the upper atmosphere showed that it was rich in carbon dioxide. Because Venus is closer to the Sun than we are, and because carbon dioxide produces a greenhouse effect, it seemed likely that Venus would be hot, but perhaps not intolerably so. Svante Arrhenius, a Swedish physicist whose work was good enough to win him a Nobel Prize, once suggested that Venus might be a world in a 'Coal Forest' condition, similar to the Earth of 225,000,000 years ago, with oceans, giant ferns, horse-tails and dragonflies flitting about, together with amphibians crawling in the swamps. This seemed rather extreme, but there was a strongly-supported theory that there might be broad oceans; if the atmospheric carbon dioxide had fouled such oceans, the rather curious result would be seas of soda-water. Other astronomers regarded Venus as a raging dust-desert, while Sir Fred Hoyle believed that there would be seas of oil. We simply did not know.

Another mystery concerned the so-called Ashen Light, or faint visibility of the night side of Venus during the crescent phase. The same effect can be seen with the Moon (look for it at almost any time when the crescent Moon is seen against a darkish sky), but this is easy to explain; it is due to light reflected on to the Moon from the Earth. With Venus, things are different, because there is no satellite. Some observers dismissed the Ashen Light as a mere contrast effect, but I did not agree, because I had seen it too often and too clearly. I did not go so far as the last-century German astronomer Franz von Paula Gruuthuisen, who believed the Light to be caused by illuminations celebrating the election of a new Venusian Parliament (!), but I did believe that it could be due to electrical disturbances in Venus' upper atmosphere, not unlike our auroræ.

Mercury is also difficult to study, partly because, like Venus, it too is 'new' when closest to us but also because it is never far from the Sun in the sky. In size and mass it is more like the Moon than like the Earth, and the low escape velocity (2.7 miles/4.3 km per second) had shown that there could not be much in the way of atmosphere.

The best chart of Mercury up to 1957 was the work of Eugenios Antoniadi, a Greek astronomer who spent most of his life in France and was able to use one of the world's best refractors, the 33-inch (83-cm) telescope at Meudon, near Paris. Antoniadi believed that the rotation period was captured, i.e., equal to the orbital period of 88 Earth-days, so that the same side of Mercury would face the Sun all the time; there would be an area of permanent daylight and another area over which the Sun would never rise, with only a narrow 'twilight zone' in between from which the Sun would bob up and down over the horizon. Science-fiction writers made great play of Mercury's twilight zone, but we now know that Mercury does not behave like this; the true rotation period is 58.6 days, or two-thirds of a Mercurian year, so that every part of the surface is in sunlight at one time or another. Antoniadi was also wrong in believing that the atmosphere was dense enough to support clouds of dust.

Before 1957 I had been able to look at Mercury through very large telescopes, including those at Paris and Flagstaff, but I had never been able to see any definite markings. It seemed reasonable to assume that the surface might not be too unlike that of the Moon, but we could not see it well enough to decide.

The giant planets were of entirely different nature. Originally they had been classed as miniature suns, and I cannot resist quoting a description of Saturn given by R. A. Proctor in the 1880s:

> over a region hundreds of thousands of square miles in extent, the glowing surface of the planet must be torn by subplanetary forces. Vast

masses of intensely hot vapour must be poured out from beneath, and, rising to enormous heights, must either sweep away the enwrapping mantle of cloud which had concealed the disturbed surface, or must itself form into a mass of cloud. . . .

It was an intriguing picture, but in the 1920s a series of brilliant papers by Sir Harold Jeffreys showed that it was wrong. The outer clouds of the giant planets are very cold indeed. In 1934 Rupert Wildt proposed a model in which a giant planet would have a rocky core, surrounded by a thick shell of ice which was in turn overlaid by a hydrogen-rich atmosphere. Hydrogen compounds, such as ammonia and methane, had been identified in the spectra of Jupiter and Saturn, and this came as no surprise; after all, hydrogen is the most plentiful substance in the entire universe. According to another model, Jupiter and Saturn at least were made up almost entirely of hydrogen, so compressed near the cores that it started to behave in the manner of a metal.

Jupiter is a fascinating object to observe, mainly because it is always changing. It spins round quickly; the rotation period at the equator is only 9 hours 51 minutes on average, but the period is rather longer in higher latitudes, and various special features have periods of their own, so that they drift around in longitude. There are belts, spots, wisps and festoons. Generally there are two main cloud-belts, one on each side of the Jovian equator, though they have been known to merge into one broad equatorial band – as I saw well for some months in 1962 – and in 1989 and again in 1991 the South Equatorial Belt virtually disappeared for a while.

Of special interest is the Great Red Spot, a huge, oval marking whose surface area is greater than that of the Earth. At its best it can be brick-red, and although it sometimes disappears for a while it always comes back. It was taken to be either a solid body floating in Jupiter's outer gas, or else the top of a column of stagnant gas. It had been under observation ever since the seventeenth century, so that it was at least very long-lived.

In 1955 it was found that Jupiter is a source of radio waves. Visual observers did their best to correlate the radio bursts with visible features, such as the Red Spot; we failed, but, surprisingly, it was found that there was a definite connection between the radio emissions and the movements of Io, Jupiter's nearest large satellite. The reason for this was unknown. We could see no definite surface details on Io or the other three large satellites, Europa, Ganymede and Callisto, but it was believed that they were likely to be ice-covered. In the case of Io, this turned out to be very wide of the mark.

Saturn is distinguished by its magnificent set of rings. No solid or liquid ring could exist – it would be promptly torn to pieces by Saturn's powerful pull – and it had been established that the rings were made up of swarms of tiny particles, presumably icy. There were two bright rings (A and B) separated by a wide gap known as Cassini's Division; close-in there was a semi-transparent ring, Ring C or the Crêpe Ring. Reports of other faint rings, outside the main system, remained unconfirmed. I remember making a careful search for them with the Lowell refractor, but I had no luck.

Like Jupiter, Saturn has a gaseous surface, and it was thought that the make-up was similar, though Saturn is considerably smaller than Jupiter as well as being less dense and less massive. Belts were visible, and there were occasional spots, though few of them lasted for long. Overall, Saturn seemed to be much less active than Jupiter. Nine satellites were known, one of which (Titan) had been found to have an appreciable atmosphere.

We did not know a great deal about the two outer giants, Uranus and Neptune, but they appeared to contain less hydrogen than Jupiter or Saturn, and rather more ammonia and water, so that the Uranus/Neptune pair differed markedly from the Jupiter/Saturn pair. Uranus was unique inasmuch as its axial tilt was more than a right angle, so that there were times when one pole remained in sunlight for a period equal to over twenty Earth-years. The reason for this was unknown – and in fact it still is. In many ways Uranus is exceptional, and unlike the other giants it does not seem to have much internal heat. There have been suggestions that early in its history it was struck by a massive body and literally knocked sideways, but this is a theory only, and I admit to being sceptical. Uranus had five known satellites, all of them smaller than our Moon, and Neptune two, one (Triton) large and the other (Nereid) small.

Pluto remained an enigma. Even its diameter was uncertain, and all we could really say was that it was almost certainly smaller than the Earth. A suggestion that it might once have been a satellite of Neptune was quite widely supported, but no telescope would show a measurable disk.

Such was the Solar System as we believed it to be in 1957. Within a few years, the whole picture had changed.

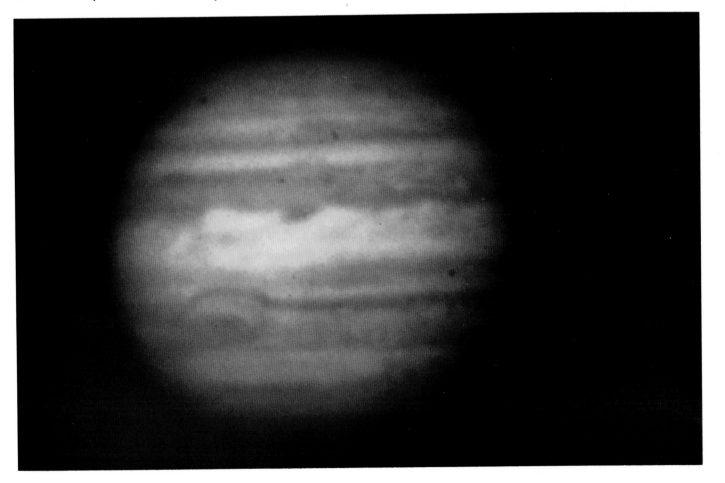

Jupiter, showing the Great Red Spot. This photograph was taken with the Palomar 200 inch (5 m) reflecting telescope. The Red Spot is to the lower left. In this picture north is at the top.

4

LUNAS, RANGERS, ORBITERS AND SURVEYORS (1958–1968)

'Men might as well try to reach the Moon as to cross the stormy North Atlantic by means of steam power.' So opined a famous scientist, Dr Dionysius Lardner, addressing the British Association in 1840. The Atlantic crossing followed soon afterwards; lunar trips were delayed for rather longer, but by the start of the Space Age, in 1957, it was clear that they would take place.

The most important problem to be cleared up concerned the possibility of deep dust-drifts. If Gold's theory had been valid, we would have had to content ourselves with orbital and unmanned vehicles. The only way to find out was to send probes there, and the first attempts were made in 1958. At this stage the American space programme was in trouble; the rockets either failed to ignite, blew up, or else plunged back to the ground in a most undignified fashion, so that it was really no surprise when the first Moon vehicles, known as Pioneers, failed. One of them (No. 3 in the series) rose to over 60,000 miles (100,000 km), which was at least encouraging, but there was still a long way to go. When success came, in January 1959, it was from the Soviet Union, with the space-craft Luna 1.

Luna 1 (or Lunik 1, as it was generally known at the time) was not designed to hit the Moon. It was a fly-by, and on 4 January it by-passed the Moon at a distance of less than 4,660 miles (7,500 km). It was a small object, only 40 inches (100 cm) across, but it was a trail-blazer, and in particular it sent back information about the Moon's magnetic field – or, rather, lack of it. It had been expected that any field would be less powerful than that of the Earth, because the smaller, less dense Moon must have less of an iron-rich core, but Luna 1 detected no field at all. We now know that there are regions on the surface which show local magnetism, and according to one plausible theory there used to be an overall magnetic field, many millions of years ago, which has now disappeared; but we cannot be sure. However, remember that if you go to the Moon you will find that your magnetic compass will not work!

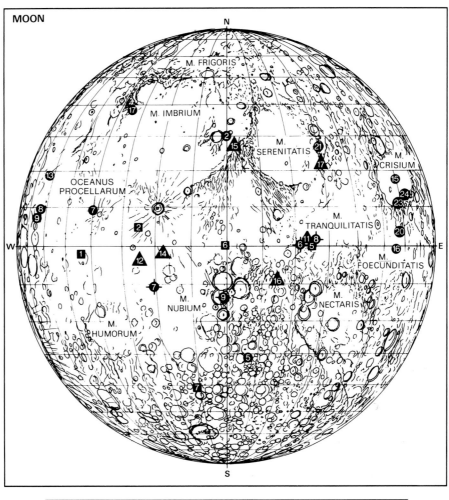

Luna ●   Surveyor ■

Ranger ⬖   Apollo ▲

Map of the near side of the Moon showing the main impact and landing sites of various Russian and American space-craft. Space-craft numbers are shown in white inside each symbol.

It is fair to say that Luna 1 did all that had been asked of it. Signals from it were lost after 34 hours, but by then its final orbit had been calculated; unless Luna 1 has been dragged off course for some reason, or has been destroyed by a collision with a meteoroid, it is now moving round the Sun in a period of 446 days, in an orbit which takes it out almost as far as Mars. In fact, it may approach Mars to within 6,000,000 miles (10,000,000 km), though unfortunately there are no Martian astronomers to observe it – at least, not yet.

The second Luna followed in September 1959. This time there was an actual landing, though the probe must have been destroyed on impact. Signals from it were picked up throughout its flight, both by the Russians and by the team of radio astronomers at Jodrell Bank in Cheshire, and these signals stopped at just the moment when an impact had been predicted. That was the sum total of Luna 2's achievements; we are not certain where it landed (probably in the grey plain of the Mare Imbrium, or Sea of Showers), but at least it has its place in history.

The real excitement came a month later. On 4 October, exactly two years after the ascent of Sputnik 1, the Russians launched Luna 3, with the stated intention of sending it right round the Moon and obtaining pictures of the hitherto-unknown far side. It was an entirely new type of experiment, and the results were eagerly awaited.

I was particularly interested, because I had spent many years in trying to chart the highly foreshortened regions round the edge of the Moon's disk as seen from Earth, and I knew that there were remarkable features there. One of these was the Mare Orientale, or Eastern Sea, which lies right at the limit of the accessible region. Years before I had been observing the area, together with my old friend H. P. Wilkins (a Civil Servant by profession, an amateur astronomer by inclination) when we had found what looked like a small sea, or mare. We reported it, drew it, photographed it, and suggested a name for it, but we had no idea then that it would prove to be a vast formation extending well over on to the Moon's hidden side; that knowledge was not to come until the mid-1960s. Meanwhile, the Soviet Academy of Sciences had asked me to send them all my observations of the foreshortened regions, and of course I had complied.

What would the far side be like? I had predicted that there would be

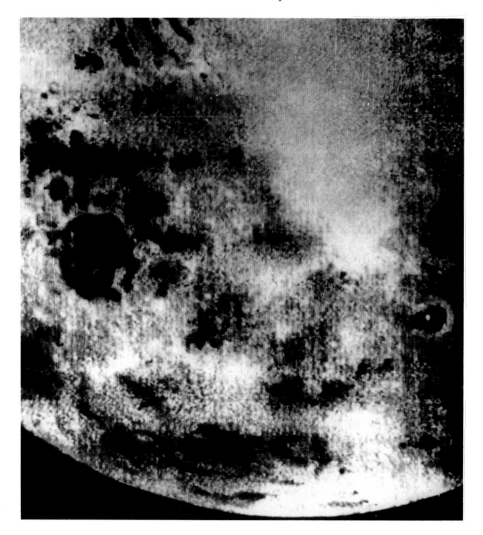

Part of the Moon, taken from Luna 3 (October 1959). The Mare Crisium is shown, as well as the area named Mare Moscoviense.

fewer maria than on the familiar hemisphere, but it seemed certain that there would be craters, mountains, ridges, valleys and all the usual features. Once Luna 3 had been successfully dispatched, there seemed every hope of finding out.

All went well. Luna 3 reached apogee, or furthest point from the Earth, on 10 October; it was then 292,000 miles (470,000 km) away from us, and was moving over the sunlit hidden regions of the Moon. Then it swung back Earthward, reaching perigee (the nearest point to the Earth; 29,000 miles/47,000 km) on 18 October. The pictures were transmitted, and six days later they were released to the world.

I had been assured that I would be among the first to receive the pictures – and I was. When they came through, I was actually in a BBC television studio, presenting a 'Sky at Night' programme. Of course it was live (there were few recordings in those days) and suddenly I heard the producer's voice in my headphones, telling me that the pictures were about to be flashed on to the screen. Frankly, I did not know what to expect, but when the first view appeared I recognized one feature, the Mare Crisium, or Sea of Crises, so that I was able to give what I hope was an intelligible commentary.

All that could really be seen were some dark patches and lighter streaks, partly because the pictures were of poor quality judged by modern standards but also because they were taken under the equivalent of full-moon lighting, so that there were no shadows. However, there was one prominent patch which looked like an immense crater. It was named Tsiolkovskii, in honour of the Russian rocket pioneer who had died in 1934, and later proved to be unique; it has a dark lava-coated floor and a high central peak, so that it seems to be something of a link between a small mare and a large crater. Several ray-craters were visible, together with a long bright streak which the Russians took to be a range of mountains; later it became clear that the feature was simply a bright ray, so that the 'Soviet Mountains' were tactfully deleted from the maps.

That, for the moment, was the end of the Russian successes. Luna 4 of 1963 was meant to make a controlled landing, but missed the Moon by 5,280 miles (8,500 km) and entered an orbit round the Sun. The scene shifted from the USSR to the USA.

American rockets had improved. The Ranger programme was put in hand; this time the aim was to crash-land probes on the Moon, and although there was no hope of survival, it was hoped that close-range pictures would be transmitted just before impact. Again there were early failures, but Ranger 7, launched on 28 July 1964, marked the opening of a new phase of lunar exploration. After a journey of 68½ hours, it came down in the Mare Nubium, or Sea of Clouds, near the ruined crater Guericke. It hit the surface at a velocity of over 5,000 mph (8,000 kph), but by then it had sent back 4,000 pictures; the last of them was transmitted only 0.19 second before impact, and showed a region measuring 105 by 150 feet (32 by 46 m), with crater-pits down to a few inches across. Rangers 8 and 9 were equally successful; they landed respectively in the Mare Tranquillitatis, or Sea of Tranquillity, and inside the walled plain Alphonsus, and the total number of photographs received exceeded 12,000.

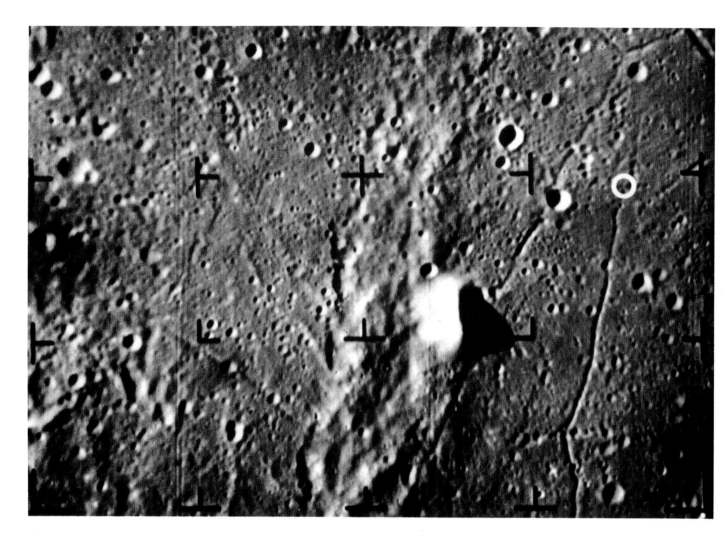

The surface did not look in the least like soft dust, and Gold's theory began to seem less and less plausible. Final proof came with Russia's Luna 9, launched on 31 January 1966. Less than a minute before impact, the rocket braking system was switched on, and Luna 9 came gently down in the Oceanus Procellarum, or Ocean of Storms. Within a matter of minutes the first pictures were being received direct from the surface of the Moon. They were picked up not only in the USSR but also at Jodrell Bank.

Luna 9 showed no sign of sinking out of sight: the deep-dust theory was dead. The scene was very like a terrestrial volcanic region, and I remember comparing it with some photographs I had taken a few years earlier in Iceland. The main obstacle to future manned landings had been removed.

Luna 9 went on transmitting for some time before its power failed. It is still standing where it came down, not far from the huge, dark-floored, walled plain Grimaldi, and it will remain there until it is collected and taken away to a lunar museum.

As so often happened at that period, a Russian advance was followed shortly afterwards by similar American achievements. Between 1966 and 1969 seven soft-landing Surveyor probes were dispatched, of which four were successful. The last Surveyor landed on the north rim of the spectacular ray-crater Tycho, which is much the most prominent feature on the Moon near full phase. As well as sending back over 50,000 pictures, the

The central peak region of Alphonsus, photographed from Ranger 9 shortly before impact (March 1965). You can see hills and the small volcanic craterlets.

Surveyors were able to analyse the material of the Moon's crust, and it was found that the surface layer was made up largely of basaltic rock, which was no surprise. However the craters had been formed, there was no longer any doubt that the Moon had been the site of tremendous volcanic activity in the past.

The White House had already announced its intention of landing a man on the Moon before 1970, but there was still a pressing need for really accurate mapping, and this led on to the Orbiter programme, which was of absolutely vital importance.

Earth-based maps of the Moon were limited, quite apart from the fact that they could reach less than 60 per cent of the total surface. What was wanted was a lunar satellite – a moon of the Moon – which would enter a closed orbit and undertake a really comprehensive survey. So, between August 1966 and August 1967, five Orbiters were sent up. All functioned perfectly, and by the time that the programme came to an end we had a really detailed knowledge of almost the whole of the Moon, including the far side. There were so many photographs that even today not all of them have been fully studied. Some of the pictures were superb; one, showing the majestic crater Copernicus, was nicknamed 'the Picture of the Century'. The programme closed on 31 January 1968, when the last Orbiter was commanded to crash itself on to the lunar surface.

The far side of the Moon was indeed rather different from the near side.

Panorama taken from Luna 9 (January 1966). This was the final proof that the Moon's surface was firm enough to bear the weight of a space-craft.

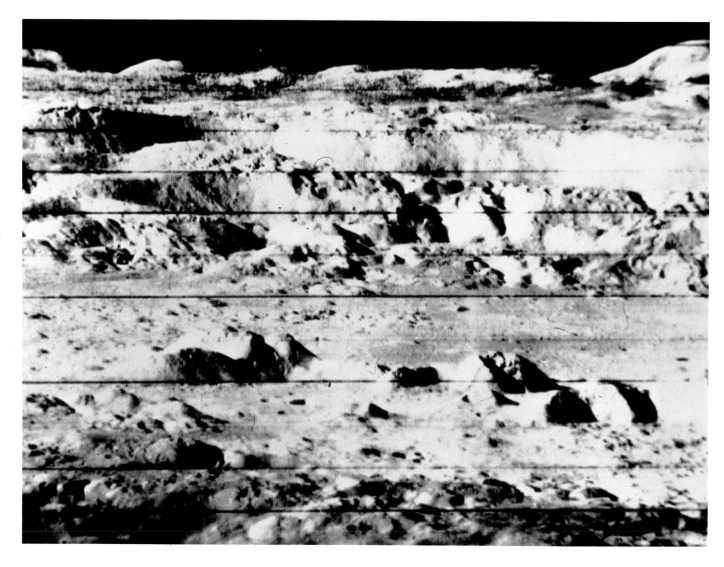

As predicted, there were no really large 'seas' apart from the Mare Orientale, which proved to be a huge, complex, mountain-ringed structure. But there were craters everywhere, together with large, light-floored enclosures which are often called thalassoids; peaks, valleys, ridges and crater-chains were very much in evidence.

Another result was the discovery of what are now termed mascons (a name derived from *mass concentrations*). As Orbiter 5 moved round the Moon, slight irregularities in its motion were detected. If a probe passes over a region where the lunar material is unusually dense, it will speed up slightly, while if it passes over an area where the material is less dense it will slow down. Dense regions were found beneath some of the regular maria, as well as the walled plain Grimaldi, which has a very dark floor and could well be classed as a minor 'sea'. It was first suggested that these mascons were buried meteorites, but this idea had to be given up when it was found that such objects would either have to be impossibly large or else much denser than iron. More probably they are volcanic, produced by the transformation of lunar basalts to denser rock at the edges of the circular structures.

The scene was set. Man was almost ready to blast off for the Moon.

The crater Copernicus, photographed from Orbiter 2 (November 1966).

5

EXPLORING THE MOON (1968–1976)

The main phase of the manned lunar programme began in 1968, when Astronauts Borman, Lovell and Anders went round the Moon in Apollo 8. They were followed in May 1969 by Apollo 10, which involved a test of the lunar module itself. The site for the first landing had been selected; it was to be in the Mare Tranquillitatis, one of the more level parts of the lunar surface, and it had been surveyed with the greatest possible care. We knew that there were no deep dust-drifts, though we could not be certain that there were no treacherous areas. We knew that the lunar atmosphere was virtually non-existent, so that there could be no radiation belts of the Van Allen type. In these respects the Moon would not make its visitors unwelcome.

Yet there was no proof that harmful materials were absent, and for this reason the pioneer astronauts were quarantined after the first landings. True, the chances of danger were very slight indeed, but they were not nil, and one could not be too careful (remember Professor Quatermass!). Quarantining was abandoned after the second mission, because by then

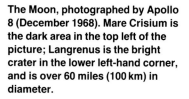

The Moon, photographed by Apollo 8 (December 1968). Mare Crisium is the dark area in the top left of the picture; Langrenus is the bright crater in the lower left-hand corner, and is over 60 miles (100 km) in diameter.

analyses had shown that the lunar rocks were sterile. Moreover, there was no trace of any hydrated materials – that is to say, materials involving the past presence of water. It was strange to recall that at a meeting of the International Astronomical Union as recently as 1966, Dr Harold Urey, one of the world's leading planetary geologists, had told me that in his view there was absolutely no doubt that the lunar seas had once been water-filled.

An Apollo crew was made up of three astronauts, two of whom went down on to the lunar surface while the third remained orbiting the Moon in the main part of the space-ship. When leaving the surface, the bottom part of the lunar module is used as a launching pad, and is left behind.

There were seven Apollo missions between July 1969 and the end of 1972, of which only one was a failure; this was Apollo 13, where an explosion in the vehicle during the outward journey meant that no landing could be attempted, and it was only by a combination of skill, courage and luck that the astronauts returned home safely. Astronauts Conrad and Bean, from Apollo 12, actually went up to the grounded probe Surveyor 3 and brought parts of it back for analysis. Rocks were collected, holes drilled to measure the outflow of heat, photographs were taken, and scientific equipment was set up, some of which went on sending back data long after the astronauts had left. By the end of the Apollo series, in December 1972, our knowledge of the Moon had been increased beyond all recognition.

I saw only one actual launch, that of Apollo 17, but throughout the Apollo programme I was carrying out commentaries for BBC television, and I have had many talks with the astronauts, so that before going on, I may perhaps be allowed to make a few personal comments based upon what I was told.

I doubt if anyone has bettered the description of the Moon given by the second Apollo astronaut, Edwin Aldrin, soon after he had followed Neil Armstrong down on to the surface: 'Magnificent desolation.' Neil's words were equally compelling:

> You generally have the impression of being on a desert-like surface, with rather light-coloured hues. Yet when you look at the material from close range, as in your hand, you find that it's really a charcoal grey. We had difficulties in perception of distance. For example, from the cockpit of the lunar module we judged our television camera to be only 50 or 60 feet away, yet we knew that we had pulled it out to the full extent of a 100-foot cable. Similarly, we had difficulty in guessing how far the hills on the horizon might be away from us. The peculiar phenomenon is the closeness of the horizon, due to the greater curvature of the Moon's surface – four times greater than on Earth; also it's an irregular surface, with crater rims overlying other crater rims.

When I asked him about the difference between the Earth-turned and the far sides, he replied: 'I would say that the striking change is in topography. There are no seas on the far side of the Moon; it's all highlands and mountains and big craters, so it's strikingly different from the side we can see from Earth.'

Walking was easy enough, and the last three Apollo crews took Lunar

Colonel Edwin ('Buzz') Aldrin on the Moon – the classic picture (Apollo 11, July 1969).

Rovers or 'moon cars' with them, so that they were able to drive around for considerable distances. However, the lunar dust was a nuisance. According to Eugene Cernan, commander of Apollo 17,

The far side of the Moon, taken from Apollo 11 (July 1969). The main crater in the picture is Dædalus, 50 miles (80 km) in diameter.

> The dust was probably the most awkward hazard of all. The dust is like graphite; but graphite lubricates, whereas lunar dust makes things stick together. It gets into your space-suits and all moving parts of your vehicles. The dust is so fine that it even gets into the pores of your skin. It took me many weeks after my return to get rid of the last traces of it.

There were various unexpected discoveries, of which the most dramatic was that of 'orange soil' in the crater nicknamed Shorty. From Mission Control, in Houston, I was watching the two Apollo 17 astronauts, Eugene Cernan and Harrison Schmitt, when Schmitt's voice came through: 'It's orange – crazy!' At first even Schmitt, the geologist of the party, believed that the colour might be due to comparatively recent volcanic activity, but when the material was analysed it was found that the colour was caused by nothing more exciting than very small, ancient glassy beads.

The lunar sky is black even in the daytime, but, to quote Cernan again:

> When you're standing in sunlight on the surface of the Moon, you can see stars if you concentrate very hard. Of course, we made several orbits of the Moon before we came in for landing, and when you're over the

Russian artist's impression of Lunokhod 1 on the surface of the Moon (January 1973).

The crater of Tsiolkovskii on the Moon's far side, photographed from Apollo 13 (April 1970). The crater seems to be intermediate in type between a crater and a sea.

dark side of the Moon you are probably in the blackest blackness anyone can imagine. You're out of sight of the Earth, and you can't even see the Moon below you. All you can see are hundreds and hundreds of stars.

I would like to spend longer in repeating what the astronauts have said to me, but this is really no place to do so, and I must also gloss over the later Russian experiments, such as the 'sample-and-return' probes (Lunas 16, 20 and 24) which landed, collected rocks and came home. There were also the Lunokhods, or 'crawlers', which looked rather like demented taxicabs but

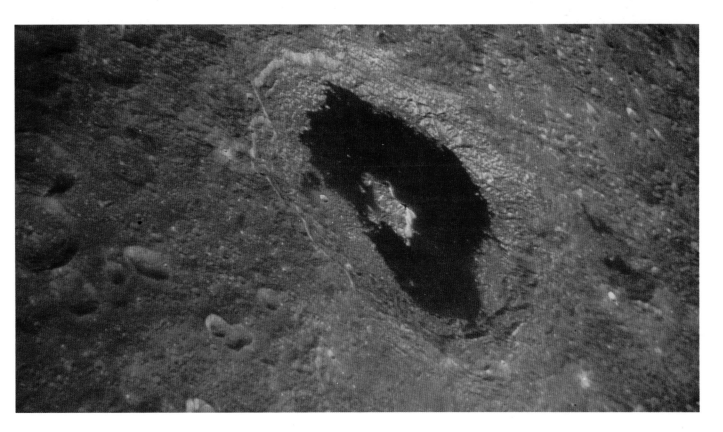

which proved to be remarkably efficient. The first of them, carried to the Mare Imbrium by Luna 17 in 1970, operated for several months; it was controlled from the ground, and sent back over 20,000 pictures. Lunokhod 2, of 1973, increased this total to 80,000. The last Soviet probe, Lunar 24, landed in the Mare Crisium in August 1976; it collected core samples from a depth down to 6 feet (2 m), and landed back on Earth on the 22nd of the month. The first phase of our exploration of the Moon was over.

The lunar surface is very old. The main Mare basins were formed between 4,000 and 3,800 million years ago, and were subsequently flooded with lava; the youngest large craters must date back at least a thousand million years, and since then there has been very little change, apart from occasional craterlets formed by meteorites crashing down on to the surface.

The surface itself is covered with an upper loose layer, termed the regolith. In places it goes down to a depth of 65 feet (20 m), though the average depth seems to be 13 to 16 feet (4 to 5 m) over the maria and about 30 feet (10 m) over the highlands. Because it is loose, it retains impressions; the footprints left by the astronauts will remain visible for a very long time, until they are either wiped out by human activity or else covered up by interplanetary 'dust'.

The rocks from the maria are basaltic. They are not unlike some of our own volcanic lavas, though they contain less silica and aluminium but considerably more iron, magnesium and titanium. The highland rocks are also igneous, with twice as much calcium and aluminium as the mare lavas, but less titanium and iron. Their age is from 4 to 4.2 thousand million years. One interesting rock-type is called KREEP, because it is rich in potassium (chemical symbol K), the rare earth elements (REE) and phosphorus (P). It also contains uranium and thorium, which are heavy, radioactive elements.

Before the Apollo missions, we did not know anything definite about the Moon's interior. Some astronomers believed it to be cold, so that the globe would be solid and inert all the way through, while others thought that it was more likely to be hot. It now seems that below the regolith there is a half-mile (1-km)-thick layer of shattered bedrock, while underneath this comes a layer of more solid rock which goes down to around 16 miles (25 km) in the mare regions. Lower down are various other layers, while the actual core, from 600 to 930 miles (1,000 to 1,500 km) in diameter, is hot enough to be molten.

On Earth, we have learned a great deal from the waves set up by earthquakes, which are of various types and behave in different ways. Moonquakes tell a similar story. They are of two sorts; some are deep-seated, 370 to 500 miles (600 to 800 km) below the surface, while others are shallow. Of course, all moonquakes are very mild compared with the shocks felt on Earth, but they do occur, and many have been recorded by the seismometers left on the surface by the Apollo astronauts. There have also been impact records, and it seems that a meteorite which hit the Moon on 17 July 1972 was at least a ton in weight.

After the last astronauts left, many of the recording stations were still in working order, and continued to send back data until September 1977, when they were shut down – partly because the equipment was starting to

Harrison Schmitt examines a large boulder (Apollo 17, December 1972).

show signs of deterioration but mainly, it must be said with regret, because of the cost of monitoring them.

What of the future? When I asked Neil Armstrong what he thought about lunar bases, he replied:

> I'm quite certain that we'll have such bases in our lifetime. Somewhat like the Antarctic stations and similar scientific outposts; continually manned, although there's certainly the problem of the environment with the vacuum and the high and low temperature of day and night. Still, in some ways it's more hospitable than the Antarctic. There are no storms,

no snow, no high winds, no unpredictable weather; as for the gravity – well, the Moon's a very pleasant kind of place to work in; better than the Earth, I think.

The Earth rising over the Moon, photographed from Apollo 17 in orbit (December 1972).

And it is fitting to give a final quote from my talks with Eugene Cernan, who is, so far, the last man on the Moon:

I believe we'll go back. We went to the Moon not initially for scientific purposes, but for national and political ones – which was just as well, because it enabled us to get the job done! When there is real motivation, for instance to use the Moon as a base for exploring other worlds in the Solar System, or to set up a full-scale scientific base, then we'll go back. There will be others who will follow in our steps.

In very recent years the Moon has not been entirely neglected. The Japanese dispatched a trial probe, Hagomoro, in 1990, and the vehicle it carried – Hiten – crashed on to the Moon in April 1993, near the crater Furnerius. Then, in January 1994, the Americans dispatched Clementine, whose main task was to complete mapping the surface – particularly in the polar regions. Clementine achieved its aim, and there have been suggestions that its equipment detected ice inside some of the polar craters, whose floors are never shadow-free and are intensely cold.

Certainly the idea of a full-scale Lunar Base is no longer far-fetched, and it will indeed be surprising if a start is not made within the next few decades. The Moon is waiting for us. And despite the Lunas, the Rangers, the Surveyors, the Orbiters and the Apollos, it has lost none of its romance or its magic.

6

THE STRANGE WORLD OF VENUS (1961–1995)

Though Venus is the nearest planet to the Earth, it is always at least a hundred times as far away as the Moon. This alone makes contact much more difficult, but to make matters worse we cannot go by the shortest route, because this would mean using much more propellant than any space-craft could possibly carry. What has to be done is to take the probe up by using a step-rocket launcher, and then slow the space-craft itself down in its orbit relative to the Earth, so that it will start to swing inward towards the Sun and will meet the orbit of Venus at a time when Venus reaches the same point. It is a complicated procedure, but it is the only one open to us at the moment. It follows that a journey to Venus will take many weeks.

Remember that at the start of the Space Age, our ignorance of the surface conditions on Venus was fairly complete; we did not know whether a probe would crash-land in a torrid desert or splash down into an ocean of water (or even oil). The Russians took the lead, and on 12 February 1961 they launched Venera 1. In view of their successes with the Lunas, there seemed every reason for optimism. Alas, it was not to be. The familiar Soviet weakness – long-range communication – soon became evident. Contact with Venera 1 was lost after a few weeks, when the probe was 4,650,000 miles (7,500,000 km) from Earth, and was never regained. It is quite likely that there was a fairly close approach to Venus around mid-May 1961, but we will never know.

The first American attempt – Mariner 1, of July 1962 – was even more disastrous. Apparently somebody forgot to feed a minus sign into the computer, and with a complicated interplanetary mission this makes quite a difference. Mariner 1 plunged into the sea and, literally as well as metaphorically, sank without trace.

It was not a good beginning, but Mariner 2, launched on 26 August 1962, more than made amends for the fate of its twin. On the following 14 December it by-passed Venus at 22,000 miles (35,000 km), and sent back invaluable data before it moved out of range and entered a permanent orbit round the Sun. Contact with it was finally lost on 4 January 1963, when it was some 54,000,000 miles (87,000,000 km) from the Earth.

Up to this time we had still been unsure of Venus' surface temperature. There had been suggestions that the apparent heat recorded by Earth-based instruments was produced in the planet's atmosphere rather than on the surface itself. Mariner 2 showed otherwise; Venus really was scorching hot, with a temperature of at least 750°F (400°C). This at once ruled out the oceanic theory, because liquid water could not possibly exist under such conditions. Svante Arrhenius' giant ferns, dragonflies and amphibians were as unreal as Lowell's canal-building Martians.

It was also confirmed that the rotation period of Venus was very long, perhaps even longer than the planet's 'year' of 224.7 days, and that Venus span in a sense opposite to that of the Earth – that is to say, from east to west instead of from west to east. Yet the French observers had already given a rotation period of only four days for the upper clouds. Could both values be right? Amazingly, this was later found to be so. The cloud-tops are moving sixty times faster than the globe, giving a rate equivalent to winds of 330 feet (100 m) per second. Mariner 2 also confirmed that the bulk of the atmosphere is made up of carbon dioxide (which had been expected) and that the clouds contained large amounts of sulphuric acid (which had not).

I am bound to admit that I found the Mariner 2 results rather disappointing. Sub-consciously, at least, I had hoped for a friendly world, but it now looked as though Venus would give any visitors a decidedly hostile reception. Mariner 5, which by-passed Venus at 2,500 miles (4,000 km) in 1967 and sent back excellent pictures of the upper clouds, told the same story.

However, we still had no direct information from below the clouds, and there were still a few people who clung to the hope of finding a surface which was not intolerably hot. Again the Russians took the lead. Bringing down a probe to a soft landing would clearly be fraught with difficulty, because everything would have to be automatic – once the landing sequence had been started, by command from Earth, it could not possibly be stopped – and the probe would have to be cooled down as much as possible before plunging into the clouds. At a conference held in Moscow, I talked to Valeri Bykovsky, a Russian cosmonaut who had been in space and knew as much about it as anybody. 'What do you think are the chances of getting data back from Venus' surface?'

Bykovsky thought hard. 'It will be a problem – much worse than with Mars – but I believe we can do it, though I am not sure that we will like what we find.'

We had not long to wait; the first partial success came with Venera 4, in 1967. The probe continued to transmit for 94 minutes after going into the clouds, but the results did not seem to fit in with the earlier data, and it was all somewhat confusing. Very naturally, the Russians had underestimated the pressure in the lower part of Venus' atmosphere, and the luckless probe had been literally squashed before it had reached ground level. As soon as this was realized, probes were constructed of tougher material. Venera 8 of 1972 did reach the surface, and transmissions went on for the best part of an hour.

I will pass briefly over Mariner 10, America's next mission, because its main target was Mercury, and it merely by-passed Venus en route, pausing to take pictures of the cloud-tops. Then, in 1975, came what was perhaps the greatest Soviet achievement to date. During October, first Venera 9 and then Venera 10 came down on to the surface, and sent back the first direct pictures. They proved to be very revealing indeed.

Mariner 2, the first successful interplanetary space-craft, which by-passed Venus in 1962.

The first real surprise was the amount of light available. It had been expected that the ever-present clouds would turn Venus into a very gloomy world, and the probes even took floodlights with them, but in fact the level of illumination was described as being about the same as that at noon in Moscow on a cloudy winter day, which, as I remember saying at the time, may not have been exactly cheerful but was a great deal better than had been anticipated. There was no evidence of any dust-cloud at the time of impact, so that Venus' regolith (if one could call it that) was presumably rigid. The area in view was rock-strewn, and there was no evidence of much erosion. This came as rather a surprise to me, because even a slow wind in that thick atmosphere would have tremendous force. We now know that there is almost no wind at all, which in view of the rapid rotation of the upper clouds means that the structure of the atmosphere must be very unusual.

The next step was to draw up a map of Venus. Early efforts had been made by Earth-based radar, and some interesting features had been found, but nobody knew quite what they were, and neither had there been any chance of analysing the surface materials. Space-craft carrying radar equipment, put into closed paths round the planet, were expected to give us the answers – or at least some of them.

America's Pioneer Venus mission, launched in 1978, involved both a radar-carrying orbiter and a multi-probe arrangement consisting of one large and three small entry probes, mounted on what was rather inelegantly termed a 'bus'. Well before Venus was reached, the entry probes were released, and on 9 December they crash-landed. Their rôle was to

Volcano on Venus; from Magellan radar results.

send back data during their descent through the atmosphere; they had no parachutes, and were not designed to survive after impact, though in the event one of them did so and went on sending back information for over an hour after arrival. The 'bus' itself had no braking mechanism, and simply burned away after plunging into the upper clouds.

The Russians kept up their efforts. Veneras 13 and 14 (1981) sent back the first colour pictures from the surface, showing rocks which glowed orange-red in the intense heat; Veneras 15 and 16 (1983) were radar mappers, while in June 1985 the two Vega probes, *en route* to Halley's Comet, dropped balloons into Venus' atmosphere and sent back data from various levels.

Since these last Russian forays, two American space-craft have encountered Venus. One of these was launched on 18 October 1989 and by-passed Venus at 9440 miles (16,000 km) on 10 February 1990, but though it

sent back some interesting pictures I will say no more about it here, because it was essentially a Jupiter probe and its meeting with Venus was merely incidental. The other, Magellan, was the most important Venus space-craft to date. It was purely a radar mapper, carrying synthetic aperture radar (SAR) capable of penetrating the clouds. It was launched from the Atlantis Shuttle on 4 May 1989; on 10 August 1990 it went into orbit round Venus, and began its main task. Closest approach to Venus was 183 miles (294 km), though its eccentric orbit, with a period of 3 h 15 m, carried it out to 5,265 miles (8,472 km). The radar covered one image 'swath' with each orbit, so that a complete mapping cycle was completed after a full Venus rotation of 243 Earth-days; this was achieved on 15 May 1991, and a second mapping cycle followed, after which the orbit was modified for gravitational measurements. Magellan finally came to the end of its career in 1994.

Quite apart from topographical work, giving a ground resolution of 400 feet (120 m), Magellan's main dish, 12 feet (3.7 m) across, sent down a pulse at an oblique angle to the space-craft, striking the surface below much as a beam of sunlight will do on Earth. The surface rocks modified the pulse before it was reflected back to the antenna, and this gave a clue as to the nature of the material, because rough areas are radar-bright while smooth areas are radar-dark. A smaller antenna sent down a vertical pulse, and the time-lapse between transmission and return gave the altitude of the surface below to an accuracy of 100 feet (30 m).

Venus is indeed a strange place. A huge rolling plain covers 65 per cent of the surface, and there are two main highland areas, Ishtar Terra and Aphrodite Terra, which are sometimes called continents, though in view of the absence of seas on Venus this is decidedly misleading. Ishtar, in the north, is 1,800 miles (2,900 km) across; its western end, Lakshmi Planum, is a lava-coated plateau, and at the east end of Ishtar lie the Maxwell Mountains, the highest on Venus, which rise to 5 miles (8 km) above the plateau. Aphrodite is larger, straddling the equator and measuring 6,000 by 2,000 miles (9,700 by 3,200 km). Adjoining it are deep valleys, one of which, Diana Chasma, is a trench over 600 miles (1,000 km) long, with a breadth of hundreds of miles and a floor 13,000 feet (4,000 m) below the nearby ridges. Diana Chasma makes our Grand Canyon look very tame!

There are large, more or less circular lowland areas; networks of faults; 'coronæ', volcanic structures up to 300 miles (400 km) across; and what are called 'arachnoids', circular volcanic structures surrounded by complex features. Craters are numerous, and lava-flows are everywhere. Of special interest are the volcanoes, of which the most notable is Theia Mons in the smaller highland area of Beta Regio; it is a massive shield volcano, close by another huge peak, Rhea Mons. Theia is probably active, and indeed vulcanism on Venus may be very widespread today. Lightning has been recorded, and we also have data showing that in 1978 the amount of sulphur dioxide in Venus' atmosphere suddenly increased by a factor of fifty, so that presumably a major eruption had sent sulphurous gases high above the ground.

Theia is by no means unique; for instance Maat Mons, slightly north of the equator, is higher and more massive than Earth's largest shield volcano, Mauna Kea in Hawaii. The Earth's crust is broken up into various 'continental plates' which move slowly around, bumping into each other and grinding against each other. When a volcano is formed over a hot spot below the crust, the volcano does not stay there; after a while it drifts away and becomes extinct. This has happened, for instance, in Hawaii. Mauna Kea, on which the great astronomical observatory has been built, has not erupted for thousands of years, and will probably never do so again (at least, one hopes not). Its neighbour Mauna Loa now lies above the hot spot, and is in a state of almost constant activity, but eventually it too will be shifted away and will cease to erupt. On Venus there are no similar moving plates, so that when a volcano is formed it stays put and has time to build up to immense size, as Theia has done.

The ground pressure of the atmosphere is about ninety times that of the Earth's air at sea-level; carbon dioxide makes up most of it, with sulphuric acid in the clouds. Since the general surface temperature has been measured as around 900°F (480°C), hot enough to melt lead, any Earth-type life can be ruled out. As I have commented, any astronaut foolish enough to go to Venus and step outside his space-craft would be promptly fried, crushed and corroded. We must be content to survey the Planet of Beauty from a respectful distance.

Why is Venus so unlike the Earth, bearing in mind that the two globes are so similar in size and mass? The answer must lie in Venus' lesser distance from the Sun. It has been suggested that in the early period of the Solar System, when the Sun was less luminous than it is now, Venus and the Earth began to evolve along similar lines, producing the same types of atmospheres and oceans, together with – perhaps – primitive life. But as the Sun became hotter, conditions changed. The Earth was far enough away to escape disaster; Venus was not. The temperature rose, the hot oceans boiled away, and the carbonates were driven out of the rocks; carbon dioxide accumulated in the atmosphere, and there was a 'runaway greenhouse' effect which transformed Venus from a potentially life-bearing world into the raging inferno of today. The water was lost; the only 'rain' on Venus now consists of sulphuric acid droplets in the upper clouds, which evaporate well before they can reach ground level.

If we could go on a trip to Venus, travelling in a space-craft which could withstand the appalling conditions, what would we see? Let us make the journey, and find out.

The top of the atmosphere lies at least 250 miles (400 km) above the surface. As we approach, the clouds below appear as dense haze; at an altitude of 44 miles (70 km) we enter the first layer, and by the time that we have fallen to 37 miles (60 km) the Sun is almost hidden, showing up as nothing more than a brilliant patch. Visibility decreases, and at 30 miles (50 km) has dropped to less than that in one of our fogs. There are alternate denser and clearer layers, but when we come to a level no more than 4½ miles (7 km) above the ground we break through into a region where clouds are almost absent; the light outside our space-craft is orange-red, and when we finally land, with an outside temperature of around 900°F (500°C), we find that the sky is itself orange. There is little wind; we are in a region of almost dead calm, but the Sun can be traced only as an ill-defined, baleful

Rocks on Venus, photographed from the Russian probe, Venera 13 (March 1982). Part of the space-craft is shown.

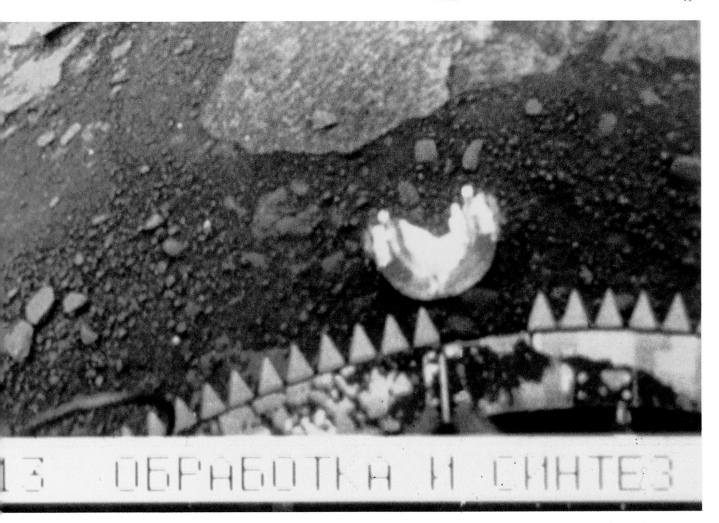

glare. The 'day' is long, because Venus spins so slowly, and we will find that the interval between one sunrise and the next is equal to more than eight Earth-weeks.

Whether this scene will actually be witnessed by Earthmen seems doubtful, and science-fiction-type ideas of 'seeding' Venus' atmosphere, breaking up the carbon dioxide and sulphuric acid to release free oxygen, seem to be well beyond our capabilities, at least in the foreseeable future. In all honesty, it has to be said that Venus has proved to be unrewarding, and the Russians announced in 1989 that henceforth they meant to turn their main attention to Mars. Where we had once hoped to find water, luxuriant vegetation and perhaps life of a reasonably advanced type, we have found instead a planet upon which conditions are remarkably like the conventional picture of hell.

7

MARINER TO
MERCURY (1973–1975)

In 1952, when I first looked at Mercury through a really large telescope, I found that I could see no markings at all. The telescope was the 33-inch (83-cm) refractor at the Observatory of Meudon, near Paris, which had been used by Eugenios Antoniadi to draw what was regarded as the best map of Mercury up to that time. I made many other observations, both at Meudon and with Lowell's telescope at Flagstaff in Arizona, but with little more success.

Like Antoniadi, I observed the planet in broad daylight, with Mercury high above the horizon. Of course the Sun was high too, but this was better than waiting until dusk, when Mercury would have been low down and its light would have had to pass through a thick, unsteady layer of the Earth's atmosphere. Eventually I thought that I could distinguish a few shadings, but with no real certainty, and I realized the immense difficulty of trying to study Mercury with any Earth-based telescope.

The first real advance in our knowledge came in 1962, when W. E. Howard and his colleagues at Michigan measured the temperature of Mercury's dark side, using infra-red techniques, and found that it was much warmer than it would be if it never received any sunlight. Evidently the rotation was not synchronous – in other words, Mercury did not keep the same face turned permanently towards the Sun, as the Moon does with respect to the Earth. Three years later, observers at Puerto Rico used the great 'dish' at Arecibo, built in a natural hollow in the ground, to make radar measurements, and found that the real rotation period is 58.6 Earth-days, or two-thirds of a Mercurian year. Every part of the planet receives its fair share of sunshine. In a way this seemed rather a pity; the idea of landing in the Twilight Zone, where the temperature would have been tolerable, had to be given up.

Earth-based radar could tell us little more, because Mercury is relatively small and is a long way away. So far as the nature of the surface was concerned, different people had different ideas. Some astronomers maintained that the surface would be as smooth as a snooker ball, because of intense solar activity in the early period of the Solar System – and remember, Mercury is close-in; its orbit is decidedly eccentric, but at its nearest the distance from the Sun is no more than 28,500,000 miles (45,900,000 km). According to another theory, the rocks would expand when heated during

The Mariner 10 space-craft. With its solar panels it has a wingspan of 25 feet (7.6 m). The camera telescopes are attached to the craft.

the long day and contract during the cold of night, making them crumble and giving the landscape a rather dilapidated appearance. My own view was that the surface was probably rather like that of the Moon, with craters, mountains and valleys, and possibly larger features resembling the lunar maria, though I doubt if anyone believed that there could ever have been oceans there.

Virtually all our information today is based on the results from one space-craft, Mariner 10, which was launched from Cape Canaveral on 3 November 1973. Mercury was the main target; *en route* Mariner passed by Venus at a distance of 3,600 miles (5,800 km), and photographed the cloud-tops, after which the gravitational pull of Venus was used to send the probe on the way to Mercury. The first pass of Mercury was made on 29 March 1974, less than two months after the rendezvous with Venus; the earliest television pictures were actually sent back on 24 March, from a distance of over 3,000,000 miles (5,000,000 km). After the encounter, Mariner entered a permanent path round the Sun, which brought it back to the neighbour-hood of Mercury on 21 September; again the television pictures were excellent. By the time of the third encounter, on 16 March 1975, the on-board equipment was showing signs of deterioration, but the results were still good. That was the end of the mission. On 24 March contact was lost forever, and to all intents and purposes Mariner was dead. It is still circling

MERCURY

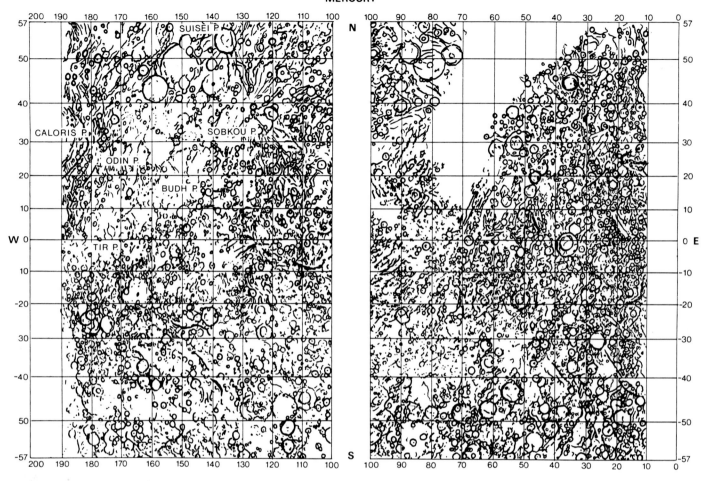

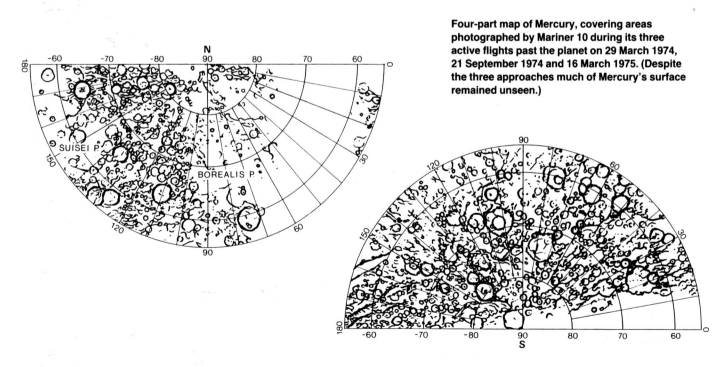

Four-part map of Mercury, covering areas photographed by Mariner 10 during its three active flights past the planet on 29 March 1974, 21 September 1974 and 16 March 1975. (Despite the three approaches much of Mercury's surface remained unseen.)

the Sun, and still making close approaches to Mercury every six months, but we have no hope of finding it again.

The main drawback was that the same part of Mercury was in sunlight at every encounter, so that we could not map the whole of the surface; we know less than half of it. However, you cannot have everything, and Mariner 10 must be ranked as an overwhelming success.

As many people had expected, the surface was cratered and mountainous. The first feature to be identified was a bright crater in the centre of a system of bright rays; it was named Kuiper in honour of Gerard Kuiper, the Dutch astronomer who had been a pioneer of planetary exploration by space-craft but who, sadly, did not live to see its full results.

As Mariner 10 closed in, more and more craters came into view. There was also one huge, mountain-ringed basin (now called the Caloris Basin) which alone rivalled the maria of the Moon, and has been compared with the Mare Orientale, though unfortunately only part of it could be seen because it was cut through by the terminator (that is to say, the boundary between the day and the night hemispheres). There were wide, smoother plains, and there were long, winding cliffs or scarps which extended for hundreds of miles. Also in evidence was a large region of hilly ground which was nicknamed 'weird terrain' because of its unusual appearance. The main difference between Mercury and the Moon was the lack of lunar-type maria; so far as could be seen, the Caloris Basin was in a class of its own.

Other problems were cleared up during the first pass of Mariner 10. There is a trace of atmosphere, but the ground pressure is only about a thousand-millionth of a millibar, so that it corresponds to what we normally call a vacuum; it is made up chiefly of helium, so that presumably it consists of helium nuclei trapped from the solar wind, and is quite different

Mercury, showing a ray-crater. The ray-craters on Mercury are very similar to their lunar counterparts.

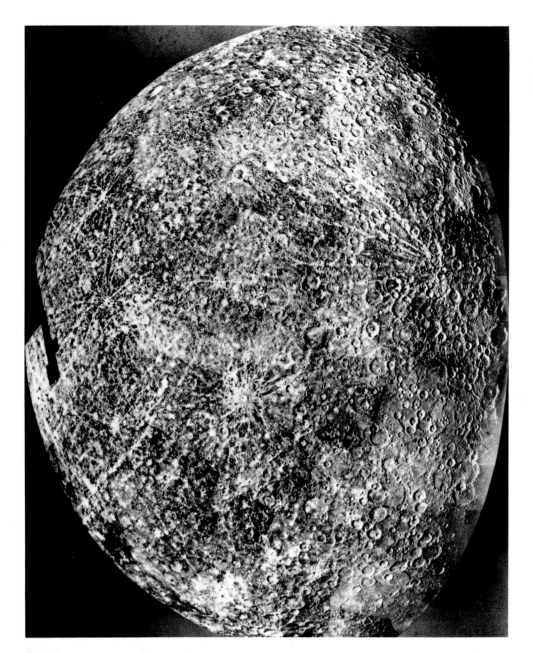

The southern hemisphere of Mercury, with its heavily cratered terrain.

from an atmosphere of the same type as ours. Certainly it is much too thin to support any dust-clouds or hazes, so that earlier reports of localized obscurations, made by Antoniadi and others, had to be ruled out of court.

More surprisingly, there was a detectable magnetic field. It is only about one-hundredth as powerful as that of the Earth, but it definitely exists, whereas Venus had shown no magnetic field at all. Apart from its weakness, Mercury's field is of the same general type as ours, and the magnetic axis is inclined to the axis of rotation by about 14 degrees.

The presence of a magnetic field, together with estimates of the planet's mass obtained from studying the movements of Mariner 10, told us a good deal about the make-up of the globe. Mercury and the Earth are the densest of all the planets, and each 'weighs' about 5½ times as much as an equal volume of water would do. It has been worked out that Mercury has a dense, iron-rich core about 2,250 miles (3,600 km) in diameter, larger than the whole body of the Moon and containing around 80 per cent of Mercury's mass.

The rotation period was confirmed, and it was also found that the rotational axis is almost perpendicular to the orbital plane. Mercury, unlike Venus, spins in the same direction as the Earth, so that to an observer on the surface the Sun would rise in an eastward direction and set toward the west. By a curious coincidence – we cannot prove that it is anything more – Mercury turns the same face towards us every time it is best placed for observation from the Earth. This was one of the facts which so misled Antoniadi and others.

The old nomenclature was abandoned, and new names were given to the features shown by Mariner 10. Craters were named after personalities, ranging from writers to artists, musicians and philosophers; among those honoured, for instance, are Beethoven, Chopin, Coleridge, Wren and Mark Twain. Valleys were named after radar installations, such as Arecibo and Goldstone; the plains ('planitia') according to the names of Mercury in different languages (such as Suisei, which is the Japanese form); ridges ('rupes') after famous ships of exploration, such as *Discovery* and *Endeavour*, and ridges ('dorsa') after astronomers who had been particularly associated with observations of Mercury, including Antoniadi. The south polar crater was named Chao Meng Fu, and it was agreed that the 20th-degree meridian should pass through the centre of a small crater, less than a mile (1.5 km) across, which lies within a degree of the Mercurian equator. This small crater was named Hun Kal, which means twenty in the language of the Maya – who used a base-20 system of mathematics.

From the Mariner results, we are now in a position to sum up the main similarities and differences between Mercury's surface and that of the

The northern limb of Mercury. Craters abound!

Moon. The heavily cratered highlands of the two bodies are of the same type; probably they are about the same age. The craters were produced in the same way, and presumably Mercury has an outer loose layer resembling the lunar regolith. Mercury has ray-craters, crater-chains and pairs and groups of craters following similar rules of distribution (for instance, when one crater breaks into another it is almost always the smaller formation which intrudes into the larger). However, Mercury shows 'intercrater plains' which are comparatively level, rolling areas with a rough texture due to large numbers of small craters, and these are not found on the Moon. Mercury also has an unique, planet-wide system of 'lobate scarps' – long, sinuous cliffs running for hundreds of miles, characterized by a rounded and lobed appearance. These too are absent from the Moon.

Much the most prominent feature anywhere on Mercury – or, rather, on that part of it shown from Mariner 10 – is the Caloris Basin, which is 800 miles (1,300 km) across, and is bounded by rings of mountain blocks rising from a mile or so (1 to 2 km) above the surrounding surface. The interior of the Basin is ridged and fractured. When it was formed, it must have affected a very wide area of the surface.

The name 'caloris' indicates heat, and the Basin can indeed become as hot as any part of Mercury. The reason for this is rather curious, and brings us on to the remarkable nature of the Mercurian calendar.

Because the orbit of the planet is decidedly eccentric, and the distance from the Sun ranges between 28,500,000 miles (45,900,000 km) and 43,300,000 miles (69,700,000 km) the apparent diameter of the Sun changes – at perihelion it is well over one degree of arc – and so does the amount of heat received on the surface. Near perihelion, an observer standing in the Caloris Basin would see the Sun near the overhead point, and the temperature would be around 800° F (430° C). This is well below the maximum

The Caloris Basin (bottom of picture), is the flatter area surrounded by mountains about 2 miles (3 km) high.

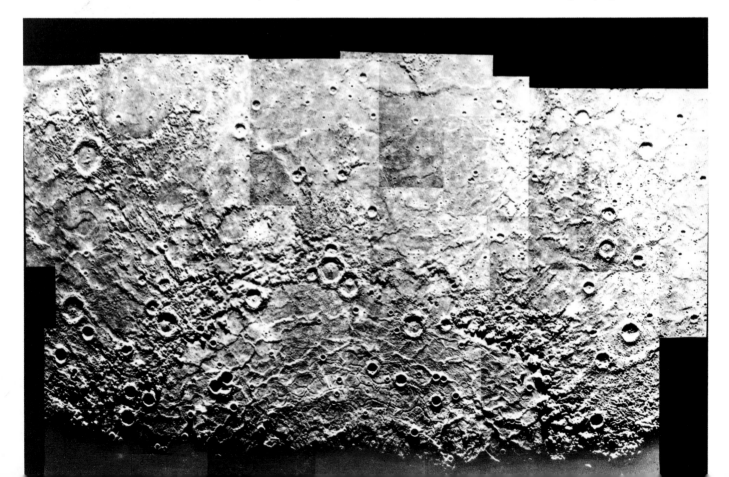

temperature of Venus, because there is no carbon-dioxide atmosphere to blanket in the solar heat; still, it is torrid enough.

At sunrise over Caloris, the Sun would be at its greatest distance. Slowly it would climb towards the zenith, growing in size; before reaching the overhead point it would stop, backtrack for a period as long as eight Earth-days, and then resume its normal motion, shrinking as it dropped in the sky and setting 88 Earth-days after having risen. This means that the interval between one sunrise and the next is 176 Earth-days, or two Mercurian years.

To an observer 90 degrees away from Caloris, the sequence of events would be different. The Sun would be largest when rising, but after having appeared over the horizon it would sink back and almost vanish before starting its upward climb to the zenith. This time there would be no overhead hovering; the size would grow again as the altitude became less, and sunset would be erratic, with the Sun disappearing, rising again briefly as though to nod 'good-bye', and then departing.

One hates to think what a 'Mercurian' would make of all this, but we can certainly rule out any form of life there, either past or present. Because no probe has actually landed, we know little about the composition of the surface rocks; it is thought that they contain less iron and titanium than those of the Moon, but unquestionably they are basaltic, and there is little doubt that Mercury has been the scene of violent volcanic activity in the remote past. Recent Earth-based radar observations have led to the claim that there may be ice inside some of the polar craters, whose floors never receive any sunlight and which are therefore intensely cold. I admit to being somewhat sceptical, and ice is surely not to be expected upon a world such as Mercury; but if it does exist, then some of our current ideas will have to be modified. Time will tell.

No further missions to Mercury are planned as yet. No doubt there will be more unmanned space vehicles during the coming century, and certainly we are anxious to map the entire surface, but at least we know that we have at last a good idea of the nature of this elusive, quick-moving little planet.

Craters on Mercury. Superficially, the surface looks very like that of the Moon, though there are differences in detail.

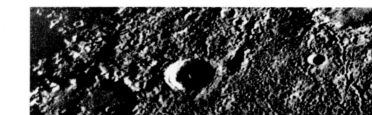

8

MESSENGERS TO MARS (1964–1993)

Ask the average non-scientist which is the most interesting of the planets, and the probable answer will be 'Mars'. This is because until less than thirty years ago it was still thought that life might exist there. Not many people (apart from flying-saucer enthusiasts) believed in 'Martians', but vegetation was another matter – and Ernst Öpik's argument, that the dark areas had to be composed of material which could push the red dust aside, was taken very seriously.

In the early 1960s I carried out an experiment together with Dr Francis Jackson, a microbiologist who is also a skilled amateur astronomer. We built a 'Martian laboratory', filled it with what we thought to be the correct atmosphere – nitrogen, with a pressure of 85 millibars – and gave it the right temperature range between day and night. When we grew things in it, the results were interesting. A cactus fared badly, and after a single Martian night looked decidedly the worse for wear, but more simple organisms did better, and we felt quite encouraged. Then came the flight of Mariner 4, and the whole picture changed.

Mariner 4 was not the first attempt. The Russians, as usual, had pioneered the experiments, but with a signal lack of success, and in fact their ill-fortune with Mars has continued right up to the present time. Their probes have either lost contact, crash-landed, or missed Mars altogether. The original American probe, Mariner 3 of November 1964, went out of control almost at once, but with its successor Mariner 4, launched on 28 November, all went well from the outset. By 14 July 1965 it was within 6,000 miles (10,000 km) of Mars, and before going on into a permanent solar orbit it had returned 21 pictures, as well as an amazing amount of miscellaneous information.

Not all this information was welcome. In particular, it was obvious that the Martian atmosphere was much less dense than had been thought. The experiment to prove this was highly ingenious. About an hour before its closest approach to Mars, Mariner went behind the planet's limb as seen from Earth; of course it was not visible, but just before and just after occultation (i.e. when it was hidden from us) its radio signals came to us via the Martian atmosphere, and were affected in a way which made it possible to work out what the atmosphere itself was like. The result was discouraging. Instead of having a ground pressure of over 80 millibars, the pressure

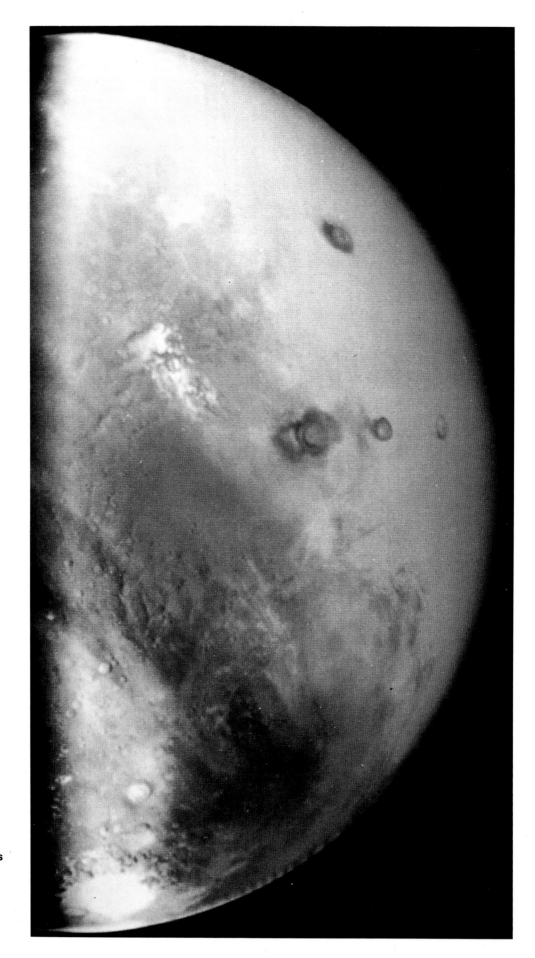

Mars, photographed from Viking 1 on its way to the planet. The Tharsis volcanoes are shown, and above them, Olympus Mons.

was very low indeed, and we now know that it is below 10 millibars. Moreover, it is not made up of nitrogen; the main constituent is carbon dioxide.

At once the chances of finding life began to recede, and it was also discovered that the dark areas are not old sea-beds; some of them, including the most conspicuous of all (the V-shaped Syrtis Major, once known as the Hourglass Sea), are high plateaux. Neither are the boundaries of the dark regions so sharp and clear-cut as they look from Earth. The regions themselves are merely 'albedo areas', where the reddish material has been scoured away by Martian winds, exposing the darker rocks underneath.

The most important revelation of all was that Mars has craters, some of them well-marked and others degraded. Evidently the planet was more like the Moon than like the Earth. Of the canals there was no sign, and neither did Mariner 4 detect anything in the way of a magnetic field.

Mariner had accomplished its task; it was a fly-by, and it could not return to the neighbourhood of Mars. Next, in 1969, came Mariners 6 and 7, which by-passed the planet a few days after Neil Armstrong had stepped out on to the surface of the Moon and, rather understandably, were somewhat over-shadowed in the popular Press. They confirmed what Mariner 4 had told

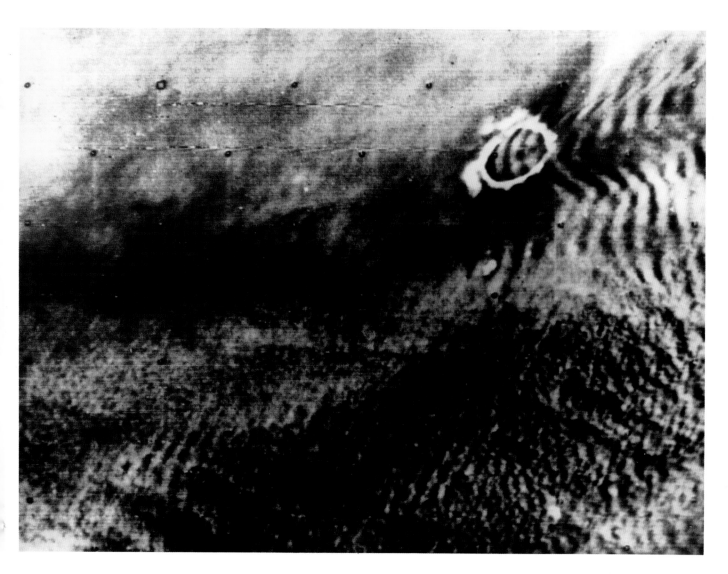

Clouds, probably of condensed carbon dioxide, with a frost-rimmed crater top about 56 miles (90 km) in diameter poking through; taken from Mariner 9.

us, and gave better views of the craters, but there seemed to be nothing startling, and it is fair to say that some of the project scientists began to feel a sense of anti-climax. I well remember a comment made to me by a leading member of the Mariner team when I was over at Mission Control in the autumn of 1969. 'We've got superb pictures,' he said. 'They're better than we could ever have hoped a few years ago – but what do they show us? A dull landscape, as dead as a dodo. There's nothing much left to find.'

How wrong he was! The really spectacular part of the programme began in 1971, when two more space-craft were sent up. Mariner 8 was a prompt failure – the second stage of its rocket launcher did not ignite, and the probe made an undignified descent into the sea – but Mariner 9, launched on 30 May, was a triumph. Instead of merely flying past its target, it was designed to enter a closed path around Mars. On 14 November its on-board motor was fired for just over a quarter of an hour, slowing Mariner down and putting it into a circum-Martian orbit which brought it down to a minimum distance of 850 miles (1,370 km) from the surface. It went on sending back pictures and data until the following 27 October, and by the end of its career it had revolutionized all our ideas about the Red Planet.

Yet at first there was little to be seen. This was something which I (and many others) had expected. Mars is subject to extensive dust-storms, which cover the whole of the surface and hide even the most familiar features; very often such a storm occurs when Mars is near perihelion, and from my modest observatory I had been keeping a careful watch. Using my 15-inch (39-cm) reflector, I had excellent views up to early October, but then the features faded out, and by the time that Mariner 9 entered its final orbit, the disk was more or less blank; in my observing book I have a comment that Mars looked 'almost like a red Venus'. Not even Mariner 9, from close range, could penetrate the dust. All that the first pictures showed were four spots which later proved to be the tops of volcanoes poking though the dust layer; we know them today as Olympus Mons (Mount Olympus), Ascræus Mons, Pavonis Mons and Arsia Mons. It was only when the dust cleared, toward the end of the year, that the real drama began to unfold. Mars was not dull; it was a world of towering volcanoes, dry river-beds, chasms and valleys, totally unlike either the Earth or the Moon. By sheer bad luck, the first three Mariners had passed over the most uninteresting parts of the entire planet.

Few people had predicted craters; almost nobody had predicted volcanoes – but there they were. Olympus Mons is three times the height of Everest, and is crowned by a huge, complex caldera. The volcano has a base 370 miles (600 km) in diameter, bounded by strange, cliff-like features,

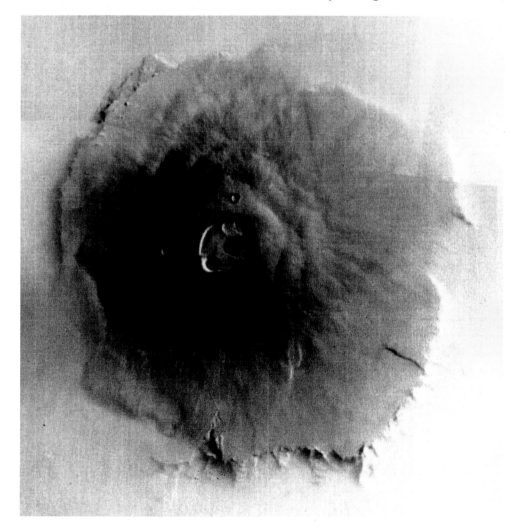

Olympus Mons, three times the height of Everest, and with a large summit caldera.

while the summit caldera is 53 miles (85 km) across. Other volcanoes, in the region known as Tharsis, were almost as impressive. All of them had been shown on earlier maps as tiny spots, and my own records tell me that I plotted them as long ago as 1939, but of course I had no idea of their nature. For example, Pavonis Mons had been known as 'Pavonis Lacus' in the old nomenclature, and had been regarded as a lake.

As Mariner 9 went round and round Mars, more and more pictures were sent back, and the final total was well over seven thousand, covering most of the planet's surface. But this was only part of the programme. The dark areas were confirmed as mere albedo features; apart from their colour, they seemed to be just the same as the reddish parts of the surface. The oval feature Hellas, south of the Syrtis Major, was found to be a basin almost 2 miles (3 km) deep rather than a lofty, snow-covered plateau. And the atmosphere was indeed painfully thin, so that it could provide no real protection against harmful short-wave radiations from space.

The canals finally passed into history. I well remember taking one of the first accurate maps and superimposing the old canal network upon it. I had rather expected to find that the sites of Lowell's canals would be marked by chains of craters, by ridges, or by differences in colour between one region and another, but I was wrong; there was no correlation at all, and it must be admitted that the entire canal system was illusory. This is forgivable when one is striving to see detail at the very limit of visibility, but the fact that so many observers 'saw' the canals, in the same places, indicates that a certain amount of unconscious prejudice must have been involved. There are wide valleys on Mars, but none of these is broad enough to be seen from Earth, and they are not connected with Lowell's canal system.

The most extensive valley complex has been named Valles Marineris, in honour of Mariner 9; it can be traced for a total length of over 2,800 miles (4,500 km), with a maximum width of 370 miles (600 km) and a depth down

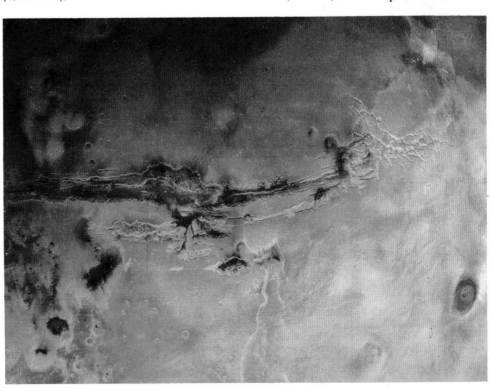

The great rift on Mars – Valles Marineris, from Viking pictures.

to 4.4 miles (7 km) below the rim. There are numerous tributaries, though the canyon system of Noctis Labyrinthus (once known as Noctis Lacus) is even more complicated, and will no doubt become a tourist attraction for future colonists.

The polar caps are of special significance in all studies of Mars. Before the flight of Mariner 9 they had been generally dismissed as very thin coatings of carbon-dioxide frost, but we now know that they are very thick, and are made up of a mixture of water ice and carbon-dioxide ice. There is plenty of H_2O on Mars, not only in the polar caps but also, without much doubt, below the surface in other areas. Liquid water cannot exist – the atmospheric pressure is too low – but Mariner 9 showed what are almost certainly old river-beds, so that water must have flowed on Mars in the remote past.

The two hemispheres of the planet are not alike. In general the southern hemisphere is the higher, more heavily cratered and presumably older, though it does contain the two deepest basins, Hellas and Argyre. The northern hemisphere is lower, less cratered and younger, though few areas are really crater-free. The boundary between these two main regions does not follow the Martian equator, but is inclined to it at an angle of over 30°; also, things are complicated by the presence of two bulges which represent the main volcanic areas. The larger of the two, Tharsis, has an area about the same as that of Africa south of the River Congo. Here we find the giant volcanoes Pavonis Mons, Arsia Mons and Ascræus Mons, with the most massive of all, Olympus Mons, not far off. There is a second bulge (Elysium) in the northern hemisphere, and also isolated volcanoes elsewhere on the surface. All the volcanoes are of the shield variety, essentially similar to those on Hawaii even though in many cases they are larger and more massive, and are associated with extensive lava-flows.

Are the Martian volcanoes active, dormant or dead? The official view is that they are extinct, but here again there may be surprises in store for us.

What Mariner 9 could not do was to decide whether or not Mars supported any form of life. The obvious presence of former rivers made the outlook a little less depressing than had been thought, but the only way to find out was to make a controlled landing on the surface – and this was achieved in 1976. Viking 1 came down on 20 July in the golden plain known as Chryse, while Viking 2 followed on 3 September in the region to the north which had been given the picturesque name of Utopia. The distance between the two landing-sites was about 4,600 miles (7,400 km).

The Vikings were twins. Each consisted of an orbiter and a lander, which crossed space together, taking over ten months to do so, and sent back splendid global views of Mars as they neared their target. Once in Martian orbit, the plan was to separate the lander, by a command sent from Earth, and bring it down gently through the Martian atmosphere, slowing it partly by parachute and partly by rocket braking. Obviously there were major problems. Once the command to separate had been sent, it could not be rescinded; everything from that moment on had to be automatic, and Mars was so far away that a radio signal took over 19 minutes to reach us. Moreover, Mars was rock-strewn, and it was impossible to tell whether or not the landing would take place on a smooth piece of ground; if the probe hit a rock it would be destroyed, while if it landed at an angle of more than a

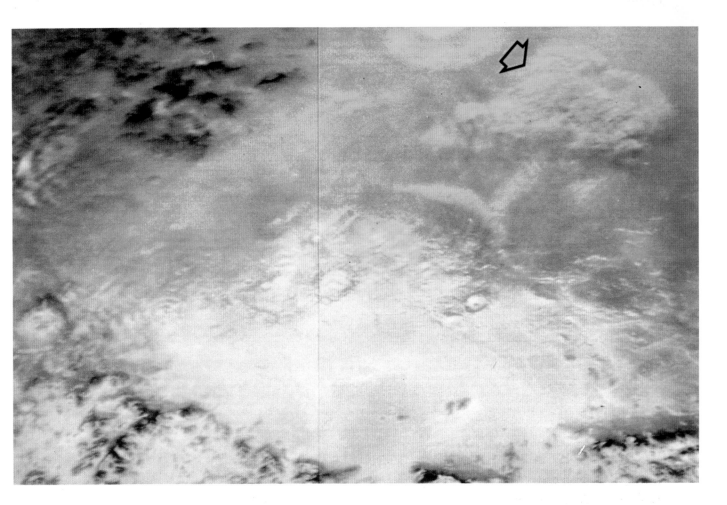

Argyre, from the orbiter of Viking 2 (22 February 1977). The arrow indicates the starting-point of a dust storm.

few degrees to the vertical it would be unable to send back signals.

During the descent of Viking 1, I was in a BBC television studio together with Dr Garry Hunt and Dr Geoffrey Eglinton, both of whom were closely involved in the Viking missions. We had some models of the space-craft, and we experimented by dropping them on to a representation of the Martian surface from various heights. I well remember Garry Hunt's wry comment: 'Well, we've proved one thing. If you drop them too hard, they break!' Actually, the landing-speed of Viking 1 was less than 6.5 mph (10 kph). The 435,000,000 mile (700,000,000 km) journey was over. Viking came down less than 26 feet (8 m) from a large boulder which would have ended the mission if the space-craft had landed on top of it.

The first picture, received immediately after touch-down, showed a rocky, red landscape, as bleak as could be imagined, under a pinkish sky. Other views followed in quick succession. The rocks were clearly volcanic; some had been smoothed by the action of wind-blown dust, while others were relatively sharp. The winds themselves were light (no more than about 14 mph/22 kph) and the climate was decidedly chilly, with a maximum temperature of about −24°F (−31°C) near noon. (Of course, Chryse is some way from the equator, and the maximum summer temperature at the equator itself rises to over 70°F (+22°C), but the thin atmosphere is very poor at retaining heat, and the night temperature must fall to around −166°F (−110°C) even at the warmest part of Mars.) New measures were made of the atmospheric composition. Over 95 per cent proved to be

The scoop from Viking Lander 1, which drew material into the space-craft and analysed it in a search for life.

Scene from the Lander of Viking 2 – a rock-strewn landscape.

View from the Lander of Viking 2 (22 February 1977). The picture shows the scoop and part of the space-craft.

carbon dioxide, with appreciable quantities of nitrogen and argon but not much of anything else. Garry Hunt gave a weather forecast from what he termed Man's most remote meteorological station: 'Fine and sunny; very cold; winds light and variable; further outlook similar.' Not surprisingly he proved to be completely accurate!

The most intriguing part of the mission was the search for life. The lander was equipped with a 'grab' capable of scooping up material from the outer surface and drawing it back into the space-craft, where it could be analysed and the results transmitted back to Earth. It was even possible to dig a shallow trench to obtain samples from below ground-level. All the three main experiments were carried through – and although the results were not so clear-cut as had been hoped, it is fair to say that no positive signs of life were found. Most people now believe that Mars is sterile, though whether any life has existed there in past ages is quite another matter.

Viking 2 followed the same programme as Viking 1. Again the landing was trouble-free, and dangerous rocks were avoided; the Utopian landscape was not basically different from that at Chryse, and all the experiments were repeated, with similar results. Meanwhile, the orbiting sections of the probes had remained in their paths round Mars, adding to the data provided by Mariner 9 and also acting as relays for the landers. Both landers and both orbiters are now dead – the orbiter of Viking 1 was the last to lose touch – but we know just where the landers are, and sooner or later we will be able to examine them.

Because of the presence of old river-beds, even with obvious 'islands', we can tell that there must have been a period when Mars was more welcoming than it is now. It has been suggested that there are temporary warmer spells when a polar cap melts, due to the changing angle of the axis of rotation, and releases quantities of water vapour into the atmosphere, and if this is true it follows that we are seeing Mars at its very worst, but at the moment we cannot be sure. Moreover, there may also be something very peculiar about Martian chemistry. We have had data from only two sites, both well away from the equator; the amount of 'soil' analysed would barely fill an ordinary test-tube; we may have looked for the wrong things, or misinterpreted what we have found. Only when we can analyse samples in our Earth laboratories will we be able to give a final answer. I very much doubt whether we will find any fossils, but I hope that I am wrong – and within a few years we ought to know.

I have not yet said much about the midget satellites, Phobos and Deimos, which look like tiny specks as seen from Earth. Both were surveyed by Mariner 9 and the Vikings, and both proved to be irregular, with darkish, cratered surfaces. Phobos has a definite regolith, and shows strange parallel grooves from 30 to 60 feet (10 to 20 m) deep; the largest crater, Stickney, is 3 miles (5 km) across – an appreciable fraction of the diameter of Phobos itself – and one cannot help feeling that if it had been formed by an impact, the whole satellite would have been in grave danger of being broken up. Deimos has a more subdued surface, with a thicker regolith and smaller craters. Neither would be of much use as a source of light during the Martian nights. They are quite unlike our massive Moon, and it is quite likely that they are ex-asteroids.

Mars, as imaged by the Hubble Space Telescope. The main feature, the V-shaped Syrtis Major, is well displayed.

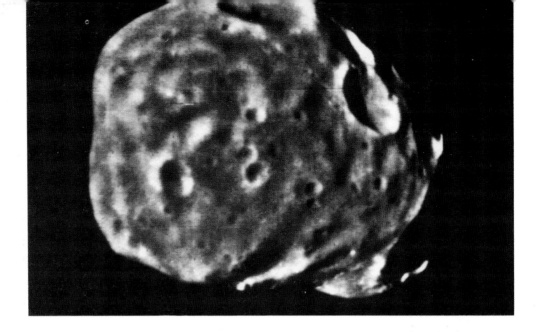

Phobos, photographed from the Russian space-craft Phobos 2 on 21 February 1989 from a range of 273 miles (440 km). Phobos, the inner and larger of Mars' two satellites, is an irregular, rocky body.

In July 1988 the Russians launched two probes, mainly to study Phobos. All sorts of elaborate experiments were planned; for instance it was hoped to fix a lander on to Phobos by a kind of harpoon, while another lander was equipped with springs which would make it capable of hopping about like a frog (remember that the escape velocity of Phobos is less than 43 feet (13 m) per second, so that a 'landing' there would really be more in the nature of a docking operation). Unfortunately, the Soviet lack of success continued. At the end of August a faulty command sent out from ground control meant that Phobos 1 lost contact, and all efforts to re-activate it proved fruitless. When I was visiting the La Silla Observatory in Chile, some months later, Dr Richard West told me that the Danish 60-inch (1.5-m) telescope there had been used to make a search for the probe, at Moscow's request, but nothing had been found. Phobos 2 did send back some pictures of the satellite, but they were not so good as the Viking views of a dozen years earlier, and in March 1989 all signals from Phobos 2 ceased because of a computer failure on the probe. Useful data were obtained, but it would be wrong to class Phobos 2 as more than a partial success.

Even this cannot be said of the latest American space-craft, Mars Observer, which was launched on 25 September 1992, and was scheduled to go into Martian orbit and undertake a very detailed mapping programme. All seemed to be well – until 25 August 1993, when contact was abruptly lost and was never regained. We will never know just what happened; we cannot be sure whether Mars Observer was put into its planned orbit, or whether it is now circling the Sun. In either case, it is dead.

Does any form of life survive on Mars? As yet we do not know; but plans for reaching the Red Planet are being drawn up. President Bush made a definite commitment to a mission there in his famous speech of 1989, which, though not so definite as President Kennedy's commitment to the Moon in 1963 (inasmuch as it gave no timescale), was at least fairly positive. At the same time I had a long talk with Colonel Yuri Romanenko, the Russian cosmonaut who had spent almost a year on the space station Mir, and he appeared very confident. It is quite possible that the first man on Mars has already been born.

9

PIONEERS AND VOYAGERS AT JUPITER (1973–1979)

Once you pass the orbit of Mars, things become much more difficult because of the vastly increased distances. Jupiter, first of the giant planets, moves round the Sun at 483,000,000 miles (778,000,000 km) on average, which is almost three and a half times the distance of Mars, so that a journey there is bound to take over a year and a half. There is also the little matter of the asteroid belt. Before the first deep-space probes were launched, nobody was at all sure how many small asteroids there might be. The main members of the swarm could be allowed for, but not the junior ones, and a collision between a space-craft and an object the size of, say, a pillow would mean disaster. Neither do the asteroids keep to the main plane of the Solar System, so that there is no way of avoiding them. Up to now, four vehicles have passed through unscathed, but it is possible that we may merely have been very lucky.

All in all, the first Jupiter probe – Pioneer 10, sent up from Cape Canaveral on 2 March 1972 – was starting upon a very hazardous and uncertain mission. It was scheduled to reach its target on 13 December of the following year; after its fly-by it would never return, but would begin a never-ending journey out of the Solar System, so that eventually its signals would fade away.

There has never been any doubt about Jupiter's status. Apart from the Sun it is obviously the major body of the Solar System, though, as we have seen, it is not nearly massive enough or hot enough to be classed as a star. The rôle of Pioneer 10 was to analyse Jupiter's atmosphere, measure the temperature, study the magnetic field, determine the nature of some of the surface features, such as the Great Red Spot, and – if possible – to send back pictures of the four large satellites Io, Europa, Ganymede and Callisto. It was certainly an ambitious programme.

The Great Red Spot was of special importance, because its origin was still uncertain – the theory that it might be a solid body floating in the Jovian gas still had many supporters. During the autumn of 1973, when Pioneer 10 was well on its way, I made many observations of Jupiter, and was relieved to see that the Red Spot was very much in evidence, in its usual latitude close to the South Temperate Belt. Pioneer came within reasonable range of Jupiter in late November, and at once it began to send back useful information.

The first revelation concerned the magnetic field. The existence of such a field was already more or less certain, if only because of the strong radio emissions, but Pioneer showed it to be more powerful than had been expected. There were also radiation belts, of the same basic nature as our Van Allen zones but far stronger. The first energetic particles were detected when Pioneer was still over 12,500,000 miles (20,000,000 km) from Jupiter, and the level of radiation increased steadily as the space-craft moved in. Radiation dosage is measured in units called 'rads'. The lethal dose for a man is 500 rads; during its fly-by Pioneer 10 received more than 250,000 rads, which shows that a manned mission to Jupiter is emphatically not to be recommended. At the minimum distance from the upper clouds (81,670 miles/131,400 km), the instruments were almost saturated, and if the approach distance had been much less the mission would have failed. The proposed orbit of the follow-up probe, Pioneer 11, was hastily modified to a different trajectory which would carry it quickly over Jupiter's equatorial zone, where the danger is at its worst.

The Great Red Spot, taken from Voyager 2, on 6 July 1979, at a range of 1,636,500 miles (2,633,000 km). Note the associated white ovals.

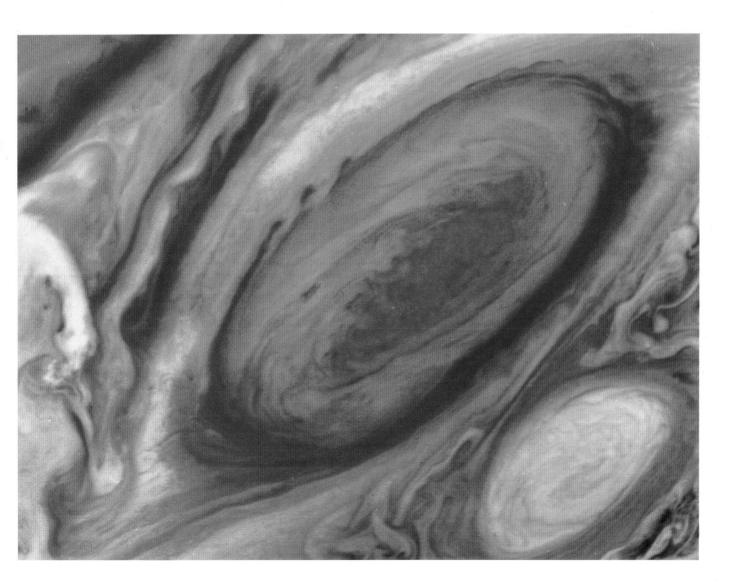

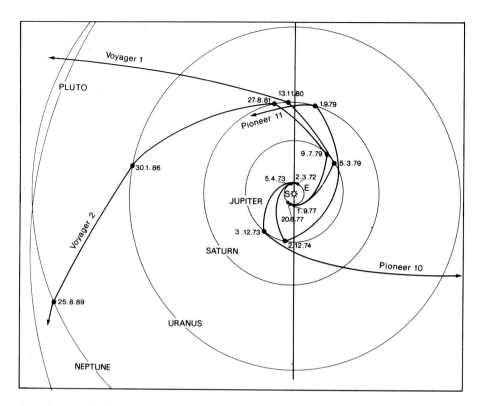

Paths of Voyager 1 and 2, Pioneer 10 and 11, with encounter dates (E = Earth, S = Sun). Pluto was not encountered by any of the spacecraft, but its orbit is included for completeness and because between 1979 and 1999 it actually comes within the orbit of Neptune.

As expected, the Jovian atmosphere was made up chiefly of hydrogen and helium, while the upper clouds were composed of ammonia droplets, and there were vivid colours. The mystery of the Red Spot was solved. The solid-body theory was wrong; so was the idea of a stagnant gas-column. Instead, the Spot turned out to be a whirling storm, spinning anti-clockwise in a 12-day period and coloured probably by phosphorus sent up from below.

Pioneer 11 followed a year later, and on 2 December 1974 it by-passed Jupiter at 28,840 miles (46,400 km), confirming the results of its predecessor. Jupiter's surface details had changed; the area round the Red Spot was different, and there were alterations also in other parts of the disk. This was not at all surprising, because Jupiter is a world in a state of constant turmoil, and the windspeeds are very high.

After Pioneer 11 had made its rendezvous with Jupiter, it still had some reserve power, and it was swung back across the Solar System to an encounter with Saturn in 1979. By then, Jupiter had been re-visited. The Voyagers, launched from Cape Canaveral in 1977, were to become the most successful deep-space probes to date. Voyager 1 by-passed Jupiter and Saturn, while Voyager 2 encountered Uranus and Neptune as well.

The multi-planet mission would have been much more difficult had it not been possible to make use of what is usually called the 'gravity-assist technique', though I rather prefer the less academic term of interplanetary snooker. By a lucky quirk of nature, it so happened that in the late 1970s the four giant planets were arranged in a sort of curve, so that it was quite practicable to send a space-craft from one to the other. Voyager 1, sent up in September 1977, by-passed Jupiter in March 1979, and Jupiter's powerful pull sent it on to its next target, Saturn, in 1980 – after which it began its never-ending journey out of the Solar System altogether. Voyager 2 was launched a few days before its twin, but travelled in a less economical path which took it past Jupiter in July 1979. Its interplanetary trek then led on to encounters with Saturn (1981), Uranus (1986) and Neptune (1989). But for the gravity-assist technique, the journey to the outer giants would have taken very much longer.

I arrived at the Jet Propulsion Laboratory at Pasadena, headquarters of the American space effort, some time before the Voyager 1 encounter. It was a great occasion, because the Voyager results were expected to be much better than those of the Pioneers. Most of the world's leading planetary scientists were present, and there was a great deal of speculation. In particular, would the Red Spot have changed much since the last close-range views of it, almost five years earlier?

We were not left in suspense for long. The pictures sent back were amazingly detailed, and there were changes both in the Red Spot and elsewhere. A thin, dark ring was detected, not in the least like the lovely ring-system of Saturn; electrical storms (auroræ) were traced over the night side of the planet, and there was evidence of majestic lightning flashes, indicating that thunderstorms were going on all the time – from close range Jupiter is a noisy planet as well as a lethal one. The Jovian magnetosphere (that is to say, the region of space over which the magnetic field is dominant) proved to be enormous, and can even extend out as far as the orbit of Saturn, so that there are times when Saturn is engulfed in it.

It was also confirmed that Jupiter sends out more energy than it would do if it depended entirely upon what it receives from the Sun. The core temperature may be at least 50,000°F (30,000°C); most of the planet is now thought to be made up of liquid hydrogen, which overlies the silicate core and is itself overlaid by the deep atmosphere.

As Voyager 1 neared its target, attention at JPL was divided between the planet itself and its family of satellites. The Pioneers had not told us much about the four senior members of the system (the 'Galileans', so called because they had been studied by the great Italian astronomer Galileo with his first telescope, in 1610), and it was hoped that Voyager would do better. The discovery of some small additional satellites was confidently expected, but in the event it was the Galileans which caused the most excitement. As the images came up on our television screens, shock followed shock.

Callisto, the outermost, is a sort of cosmic museum. The surface is crowded with craters, and is icy; there are also two large, ringed basins, now named Valhalla and Asgard, which are very striking indeed. Valhalla, 370 miles (600 km) across, is surrounded by concentric rings. Callisto is utterly inert, and one has the impression that absolutely nothing has happened there for thousands of millions of years. Ice must make up a large percentage of the globe, but it is quite likely that below the crust there is a layer of liquid water surrounding the core.

Next comes Ganymede, which is the largest satellite in the Solar System and is actually larger than the planet Mercury, though it is less massive and has no detectable atmosphere. Like Callisto, it is icy and crater-scarred, but there are also strange grooves and furrows, plus a large, darkish area which has been suitably christened Galileo Regio. Also like Callisto, it is inert, but there is evidence of crustal movements in the distant past.

Europa was completely different – so different, indeed, that when I saw the first picture of it I thought that there must be some mistake. (One eminent astronomer, who was with me at the time, commented that 'there sure must be something wrong with that camera.') There were no craters at all; all we could see were dark lines, representing features only a few feet deep and giving the surface rather the look of a cracked eggshell. Europa is a map-maker's nightmare, because one part of it seems so like another. There is an almost complete absence of vertical relief.

Later came a remarkable theory which is worth mentioning, though I am bound to say that I take it with a very large grain of cosmic salt. It has been suggested that below the icy surface there is an ocean of ordinary water; when cracks open in the crust, as must sometimes happen, a certain amount of warmth percolates inwards, and the water is kept at a temperature high enough to support primitive life-forms. It is indeed hard to visualize what existence would be like in a pitch-dark, utterly calm underground ocean!

Just why Europa is like this remains a mystery. If craters ever existed, they must have been obliterated by soft ice welling up from inside the globe, but it is all very peculiar.

Even Europa's curious aspect pales compared with that of Io, the inner-

The ringed basin Valhalla, on Callisto, taken from Voyager 2. The central area of Valhalla has a diameter of 190 miles (300 km).

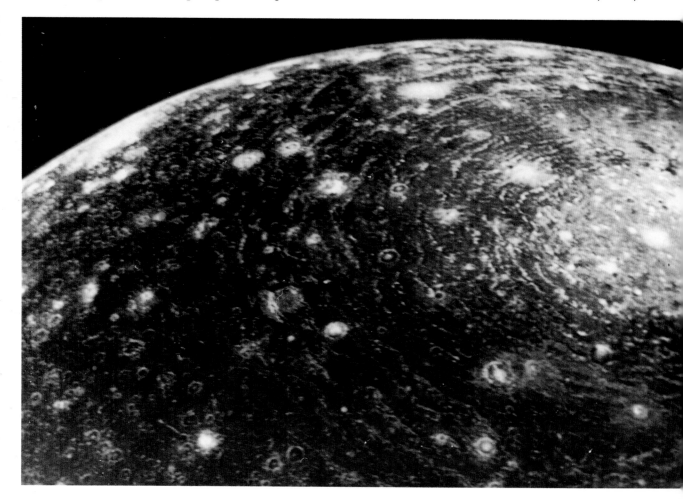

Ganymede from Voyager 1, on 5 March 1979, from 155,500 miles (250,000 km). The resolution is down to 2.8 miles (4.5 km). Note the bluish ray craters.

Europa from Voyager 2, on 9 July 1979, from 150,000 miles (241,000 km). The surface is strikingly different from that of the other Galileans.

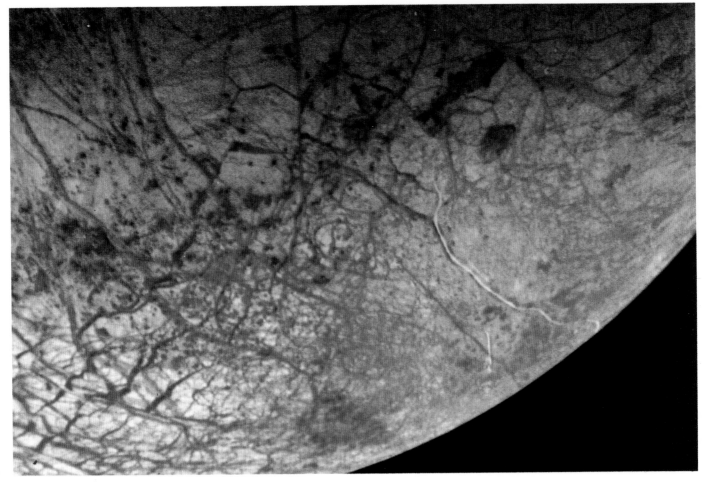

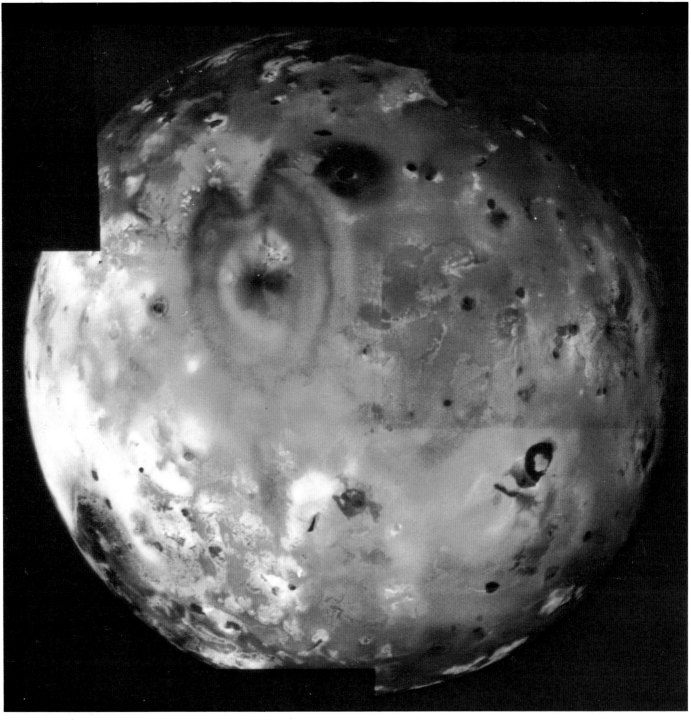

most Galilean, which is very slightly larger than our Moon. Looking at the Voyager pictures, it is easy to see why so many people likened Io to an Italian pizza. The surface is red and sulphur-coated, with yellow, brown, black and white patches. There are violently active sulphur volcanoes. During the Voyager 1 encounter, eight 'plumes' were found to be sending material up to heights of hundreds of miles; subsequently the volcanoes were named after deities associated with fire – Pele, Marduk, Loki and so on. Though the general surface temperature is very low, around −230°F (−145°C) during the daytime, the hot volcanic vents can reach several

Colour picture of Io; a computer-generated mosaic (Voyager 1). The resolution is down to 5 miles (8 km).

hundreds of degrees. There are lava-lakes, crusted lava regions, and features which are almost certainly pools of liquid sulphur. The whole surface is unstable, and to make matters even more hazardous Io moves right inside Jupiter's main radiation zone.

During the Voyager 1 encounter, Pele was the most active of the 'plumes'. By the pass of Voyager 2, four months later, Pele had ceased to erupt, but there is no reason to believe that it is extinct – it can hardly have put on a special display for Voyager 1's benefit. Moreover, some of the other volcanoes, such as Loki, were more violent during the Voyager 2 rendezvous than they had been earlier.

Io must be hot inside, and the favoured theory is that the interior is being constantly flexed by the changing gravitational pulls of Jupiter and Europa, though it is not easy to see why Io is so active and Europa so inert. Io's molten silicate core may be overlaid by a layer which is in turn overlaid by a sea of molten sulphur several miles deep, with only its outermost part in a solid state. Liquid sulphur thrusts up towards the surface and breaks through, so that the expanding gas explodes into space. Activity goes on all the time, and must presumably have done so for thousands of millions of years.

It was already known that Io's position in orbit has a marked effect upon the radio emissions from Jupiter. The Voyagers showed that the two bodies are connected by a strong electrical current, and that material ejected from the Ionian volcanoes is put into orbit round Jupiter, so that the result is a sort of doughnut-shaped 'torus' with Io in the middle.

To me, at least, Io represented the highlight of the Voyager 1 encounter. The volcanoes had been so unexpected, and it took time for us to appreciate how widespread their effects are; they may even have contaminated the smaller inner satellite, Amalthea, which has a reddish surface with craters, ridges and troughs.

Voyager 2 followed its twin, and on 9 July 1979 passed Jupiter at 443,750 miles (714,000 km), confirming the earlier findings. Since then two further probes have been sent to the Giant Planet. Ulysses was launched on 6 October 1990; its aim was to study the poles of the Sun, which can never be well observed from Earth, or from most satellites, because we always see the Sun more or less broadside-on. To be put into a path well out of the ecliptic, Ulysses had first to bypass Jupiter and make use of the powerful Jovian gravity; on 8 February 1992 it skimmed past Jupiter at 235,000 miles (378,000 km). It was not designed to make observations of Jupiter, but the chance was really too good to be missed, and data were obtained about the Jovian magnetosphere, radiation zones and general environment.

Galileo, launched on 18 October 1989, was purely a Jupiter mission, though during its somewhat tortuous path it had to make passes of Venus (10 February 1990) and the Earth twice (in December 1990 and December 1992). It consists of an orbiter and an entry probe. On arrival at Jupiter, in late 1995, the plan is to send the entry probe down into the clouds, so that it will send back useful data before being destroyed; the orbiter will undertake a prolonged 'satellite tour', making close approaches to all the Galileans. Unfortunately the high-gain antenna has failed to deploy, and this is bound to affect Galileo's performance, but the probe has already sent back two excellent views of asteroids, 951 Gaspra and 243 Ida. Gaspra was imaged on 13 November 1991, from 9,900 miles (16,000 km) and found to be small, irregular and cratered. Ida is rather larger, measuring 35 × 15 × 15 miles (56 × 24 × 24 km); it was imaged on 28 August 1993, and it also showed a cratered surface. It is accompanied by a 1 mile (1.6 km) satellite, Dactyl, at a separation of about 60 miles (100 km).

However, the latest revelations about Jupiter do not come from spacecraft at all. In July 1994 the Giant Planet was hit by a comet.

The story began on 24 March 1993, when three American astronomers, two professional (Eugene and Carolyn Shoemaker) and one amateur (David Levy) discovered what they described as a 'squashed comet', shown on a plate taken with the 18 inch (0.46 m) Schmidt telescope at Palomar. The comet was of the fourteenth magnitude, and there was something very unusual about it; apparently it had been broken up. Calculations showed that it had probably been orbiting Jupiter for at least twenty years, and that on 7 July 1992 it had approached the planet to within 75,000 miles (120,000 km), suffering badly in the process. It was still orbiting Jupiter, but its 'apojove' was some 30,000,000 miles (48,000,000 km) from the planet, and it had now moved into a collision course. Between 16 and 22 July 1994 it was clear that the comet would crash to destruction into Jupiter's clouds.

Once the announcement was made, popular interest was intense, and there were the usual alarmist reports; one earnest lady, who claimed to be Polish, took advertisements in national newspapers, confusing Comet

Asteroid Ida, as imaged by the Galileo space-craft; the tiny satellite Dactyl is shown to the right (28 August 1993).

The broken comet Shoemaker-Levy 9, photographed by David Jewitt with the 88 in (220 cm) telescope on Mauna Kea. The nucleus has been broken into a 'string of beads'.

Shoemaker-Levy with Halley's Comet, and predicting that the world would come to an end unless we took suitable precautions (notably giving up alcohol, and destroying all nuclear weapons). Of course there could be no effect upon anything other than Jupiter and the luckless comet; even the effects on Jupiter would be minor (on television, I commented that it was rather like trying to divert a charging rhinoceros by throwing a baked bean at it). But nobody knew quite what would happen, and as July drew near many large telescopes were aimed at Jupiter. It was known that the actual impacts would occur on the side of the planet turned away from us, but Jupiter's quick spin would soon bring the affected areas into view.

I was able to use the 26½ inch (67 cm) refractor at Herstmonceux, the old site of the Royal Greenwich Observatory; the telescope had been mothballed for years, but we were able to renovate it just in time. The pieces of the shattered comet hit Jupiter in turn, and the effects on the cloud-deck were much greater than had been expected, so that I was frankly staggered by the size and intensity of the storms. Predictably, the best pictures of all were obtained from the Hubble Space Telescope. The largest fragment of all, lettered G, impacted on 18 July; the icy body was a few miles across, and came in at a 45-degree angle, making a disturbance with a maximum diameter of well over 7,000 miles (12,000 km).

Inevitably it was suggested that the Earth, too, could be hit by a body of this sort, and we have to admit that such an event is possible; as we have noted, many people believe that it was a comet or asteroid impact, 65

million years ago, which changed our climate to such an extent that the dinosaurs died out. I am decidedly sceptical about this, and even if such a collision did occur I have no doubt that we would cope with the situation better than the dinosaurs are supposed to have done, but compared with Jupiter the Earth is a very small target.

We may hope for useful data from the Galileo probe, despite its faulty antenna, and moreover the Hubble Space Telescope has proved capable of monitoring changes on the planet as well as the volcanoes on Io. So Jupiter remains under surveillance, and it is a source of never-ending interest.

Impact sites of the collision with Jupiter; eight sites, fragments E/F, H (star), N, Q1, Q2, R, D/G. The D/G complex also shows extended haze at the limb. Resolution down to 124 miles (200 km).

10

SATURN (1979–1981)

When I was a small boy (and that is a very long time ago now) I had my first view of Saturn through a telescope. I remember giving a shout of wonder. I thought then, and still do, that Saturn is the most beautiful object in the sky. Other planets have rings, as we have found, but none can compare with the glory of Saturn's system.

Half a century later I had another 'first' view of Saturn, this time on a television screen at the Jet Propulsion Laboratory in Pasadena. Voyager 1 was closing in, and had already started to send back pictures far surpassing any I had seen before. True, Pioneer 11 had by-passed Saturn some months

Saturn taken from Pioneer 11 on 31 August 1979, from 586,000 miles (943,000 km). The F-Ring is shown, together with two satellites – Janus and Tethys.

earlier, but so far as Pioneer was concerned Saturn was no more than an afterthought, though it had shown enough to whet our appetites.

Saturn's globe had been expected to be less active than that of Jupiter, and there was nothing comparable with the Great Red Spot, but there were cloud belts, wisps, festoons and bright zones. What we had learned about the make-up of Jupiter also applied, with some modifications, to Saturn; it was assumed that there was a silicate core, surrounded by layers of liquid hydrogen and then the gaseous atmosphere. Saturn, like Jupiter, has an internal heat-source, and there was thought to be a magnetic field, though neither the field nor the associated radiation zones were likely to be as strong as Jupiter's.

The satellite systems of the two giants are not alike. Jupiter has four large attendants (the Galileans) and a dozen small ones, of which the outer four have retrograde motion and are probably captured asteroids. Saturn has one really major satellite, Titan, which up to the Voyager encounter was thought to be larger even than Ganymede; actually it is slightly smaller, but, unlike any of the Galileans, it has an appreciable atmosphere. Saturn's retinue also includes eight other satellites with diameters between 125 and 1,000 miles (200 and 1,600 km), and a few smaller members, though only one satellite – Phœbe, the outermost – has retrograde motion and is presumably asteroidal.

It was anticipated that Voyager would reveal some new small satellites, and there was also a chance that it would extend the known ring system. An extra inner ring had been reported, almost touching the cloud-tops; I

Saturn from Voyager 2, on 4 August 1981, from 13,000,000 miles (21,000,000 km). Four satellites are shown: Tethys, Dione, and Rhea (left) and Mimas (a bright spot on Saturn's limb above Tethys). The shadow of Mimas is visible on the planet's disk immediately above that of Tethys.

had looked for it with powerful telescopes without success, and I was frankly sceptical, but I was quite ready to be convinced. There was rather more evidence for new faint rings outside the main system. The rings themselves were expected to be more or less uniform, and to be no more than a few miles thick at most. The Cassini Division was assumed to be empty, and at one stage there had even been talk of sending Pioneer 11 straight through it. Other divisions in the rings were regarded as dubious; Gerard Kuiper, observing with the Palomar 200-inch (5-m) reflector years before, had dismissed them as 'ripples'.

By 6 November 1980, Voyager 1 was within about 6,000,000 miles (10,000,000 km) of Saturn, and already it was clear that the ring system was more complex than anyone had believed. Frankly, we were all taken by surprise. The rings were divided into large numbers of narrow ringlets and minor divisions, so that, as one NASA scientist commented, they had 'more grooves than any gramophone record'. As the view became plainer and plainer, the number of divisions rose to thousands. At one point a star passed behind the rings, and its rapid 'winking' on and off as it was hidden and revealed in quick succession provided extra proof.

Moreover, Ring B, the brightest of the rings, showed dark radial spokes which persisted for hours after emerging from the shadow of the globe. When they disappeared, they were replaced by new ones. Following the usual traffic laws of the Solar System, the ring particles closest to the planet must move more quickly than those which are further out, so that no radial features should form – yet they were quite unmistakable. My immediate reaction was that they were due to particles elevated away from the main ring-plane by electrostatic or magnetic forces, and this now seems to be the usually-accepted explanation, though it is by no means entirely satisfactory. When Voyager passed 'below' the rings (that is to say, on the side of the rings opposite to the Sun) the spokes appeared bright instead of dark, so that presumably the sunlight was being scattered forwards.

Next, it was found that two rings, one closer-in than the Crêpe Ring and the other in a dark gap near the outer edge of the Cassini Division, were not circular; they were definitely eccentric. The Cassini Division, 2,500 miles (4,000 km) wide, was not empty at all. Outside the main system came the dim Ring F, which caused more argument than anything else up to that time; it looked as though it were 'braided', or made up of several components which were intertwined. 'It breaks all the laws of nature', was one comment. It was also found that the Ring-F particles are kept in place by two small satellites (Prometheus and Pandora) which move to either side of the ring – Prometheus a little closer in, Pandora a little further out. Prometheus is moving slightly faster than the ring particles, so that it will tend to speed up a particle which strays away from the main region; Pandora, moving at a slightly lower velocity, will slow an errant particle down, so that in both cases the particles will be returned to their proper paths. These two dwarf moons, both less than 100 miles (150 km) in diameter, are appropriately known as shepherds. Other small shepherd satellites were looked for inside the main system, but only one was found; this is Atlas (less than 25 miles/40 km across) which moves just beyond the outer edge of the bright Ring A. Much later a tiny satellite, now named Pan, was identified inside the Encke division.

The ring temperatures were found to range between −290°F (−180°C) in sunlight down to −330°F (−200°C) in shadow. The sizes of the icy particles were given as anything between pebbles and pillows, but even Voyager could not show them independently. The ring thickness was reduced to no more than 500 feet (150 m), and perhaps less. Two new outer rings were found (G and E), one of which extends out to the orbit of the satellite Enceladus, but I had been right in thinking that there was no inner ring reaching down to the cloud-tops, even though there must be a good many particles there.

Even today we are not sure about the dynamics of the ring system. The old theory, according to which the Cassini Division and the other minor gaps were due to the gravitational effects of Saturn's satellites, has had to be modified because there are far too many narrow divisions, though no doubt the satellites play an important rôle. I will never forget my first sight of the rings in all their splendour. It looked to me more like an artist's impression than a true image sent back from so many millions of miles away in space.

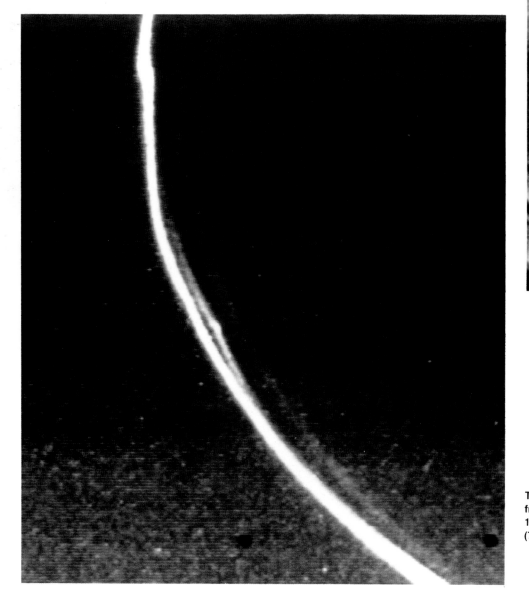

The braided F-Ring, photographed from Voyager 1, on 12 November 1980, from 466,000 miles (750,000 km).

False-colour view of the rings, taken from Voyager 2, from 55,300,000 miles (8,900,000 km). Note the difference in colour between Ring C (blue in this picture) and the rest.

A magnetic field was detected – a thousand times stronger than ours, though much weaker than Jupiter's – and it was also found that the magnetic axis and the rotational axis are almost coincident. On the disk, violent wind speeds were recorded, and there was one jet blowing at a full 900 mph (1,400 kph), which is faster than anything on Jupiter. Strangely, the boundaries between the different wind zones do not follow the edges of the belts, as on Jupiter. Saturn's spots were smaller; one red oval was detected from Voyager 1, and was still in existence at the time of the Voyager 2 pass nine months later, but whether or not it will be really long-lived remains to be seen. All in all, Saturn's disk was much blander than that of Jupiter, and the colours were less vivid. The lower temperature as compared with Jupiter means that ammonia crystals can form at higher levels, and produce a kind of uniform haze.

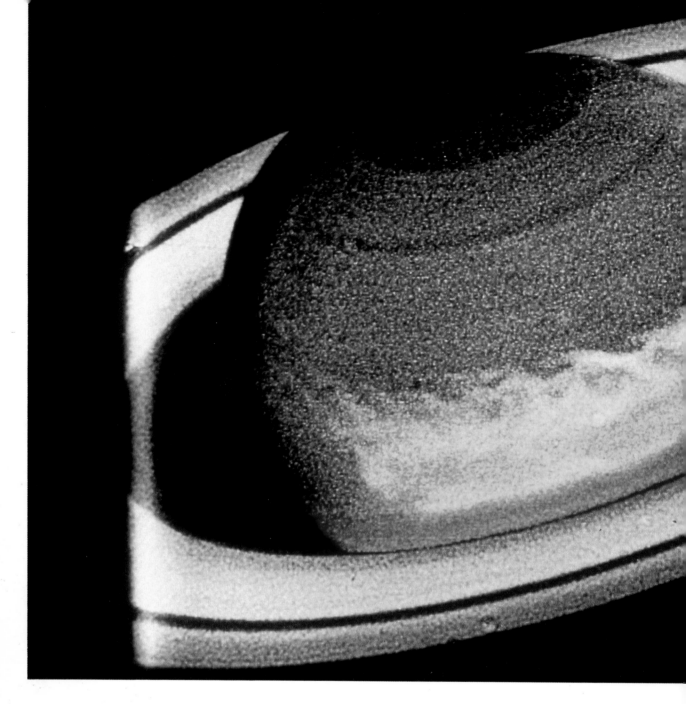

Of course Saturn itself was the main target for Voyager 1, but Titan, the senior satellite, was also of special interest, because it was the only planetary satellite known to have an atmosphere. Way back in 1944 Gerard Kuiper had used spectroscopic observations to show than an atmosphere existed, but we did not know much about it, and it was generally thought to be rather thin, with methane dominant. We hoped that Voyager 1 would tell us.

However, there was one potentially unfortunate complication. If Voyager were programmed to fly close to Titan, it would move out of the ecliptic or main plane of the Solar System, so that it would be unable to go on to the outer giants Uranus and Neptune. It was a straightforward choice. The planners decided to opt for Titan. If Voyager 1 had failed, then Voyager 2, already on its way, would have had to survey Titan, and Uranus

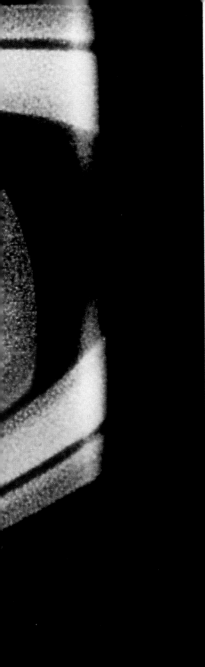

and Neptune would have been abandoned altogether, which would have been a tremendous loss.

Titan was definitely an enigma. Early images of it showed nothing but an orange, featureless disk which did not look at all promising. I remember a talk I had with Garry Hunt a few hours before closest approach; I thought that we would see the surface details through a thin methane haze, while Garry did not, maintaining that we would be able to make out nothing more than the top of a layer of what might be termed 'smog'. As Voyager 1 passed by Titan, at only 4,030 miles (6,490 km), it became clear that Garry had been right. Titan was almost plain orange, with the northern hemisphere slightly the darker of the two and with some high-altitude haze.

In spite of this, we learned a great deal. The most unexpected discovery was that the atmosphere is dense, with a ground pressure about one and a half times that of our air at sea-level, and it is made up chiefly of nitrogen. Remember, nitrogen makes up 78 per cent of the air that you and I are breathing, but we must not take the similarity too far; on Earth, most of the rest of the atmosphere is oxygen, whereas on Titan the other plentiful gas is methane. Titan itself is about twice as dense as water, so that the globe presumably consists of about 55 per cent rock and the rest ice.

The main objection to life on Titan is the very low temperature, which was measured at about −290°F (−180°C). This is near the 'triple point' of

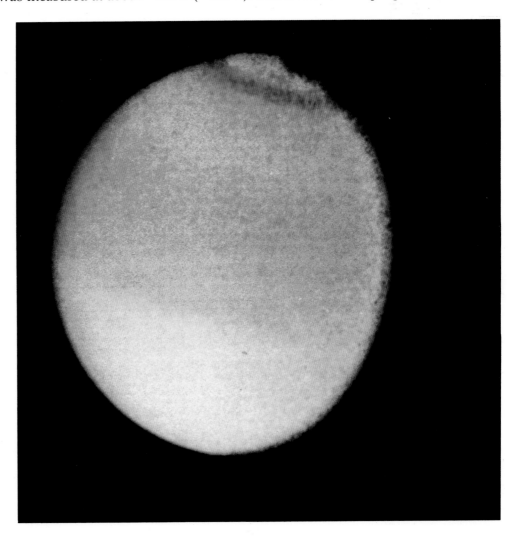

Titan, taken from Voyager 2, on 23 August 1981, from 1,400,000 miles (2,300,000 km). The southern hemisphere is light, with a band near the equator and a collar round the north pole.

methane, so that methane can exist as a gas, a solid or a liquid, just as H_2O can do on Earth as water vapour, ice or liquid water. Titan may have oceans of liquid methane, with cliffs of solid methane and a methane rain dripping down from the orange clouds in the nitrogen sky. In any case, it is quite unlike any other world in the Solar System.

Good views were obtained of most of the other satellites, apart from Phœbe. Most of them proved to be icy and cratered, though one – Enceladus – was unexpectedly smooth. Two of the new satellites, Janus and Epimetheus, share the same orbit, and every four years they exchange places in a sort of cosmic game, so that almost certainly they represent fragments of a former larger object which broke up; both are irregular in shape. (I may add that in 1966, when I was observing Saturn from the Armagh Observatory, I certainly recorded one or both of these tiny satellites, but since I did not recognize them as being new I can claim absolutely no credit.) Tethys, which had been known about ever since 1684, was found to have two smaller satellites sharing its orbit, while Dione has one 'co-orbital'.

As Voyager 1 drew away from Saturn, we were left with plenty of food for thought. Less than a year later we were back, this time for the pass of Voyager 2. All the previous results were confirmed, and surveys were made of the satellites which had not been adequately covered earlier; of these, Tethys was found to be almost pure ice, with a huge trench stretch-

Enceladus, from Voyager 2, on 25 August 1981, from 74,000 miles (119,000 km). Resolution on this picture is down to 1.2 miles (2 km). (Left)

Dione, photographed from Voyager I, showing the cratered landscape; the most prominent craters are Dido and Æneas. (Right)

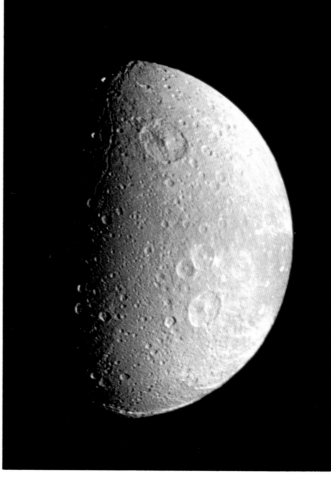

Iapetus, photographed from Voyager 2, on 22 August 1981, from 684,000 miles (1,100,000 km). The resolution is down to 13 miles (21 km).

ing for over 1,200 miles (2,000 km) around it, while the smaller Hyperion is shaped rather like a hamburger. Iapetus, 892 miles (1,436 km) across, has one hemisphere which is as bright as ice and the other which is as black as a blackboard, so that presumably dark material has welled up from below the crust, covering a large part of the globe.

There was a brief alarm when the signals from Voyager 2 were interrupted during the outward journey, but the trouble did not prove to be really serious, and as the space-craft went on its way we all looked forward to the encounter with Uranus in January 1986.

11

VOYAGER 2 AT URANUS (1986)

By the time that Voyager 2 started to draw away from Saturn, it had already proved itself to be one of the most successful space-craft ever launched. Despite the temporary loss of pictures at the very end of the Saturn encounter it had carried out all its tasks, and in many ways it had improved upon the performance of its twin. And unlike Voyager 1, it had more to do. Its next target was Uranus, the strange, green planet which had been discovered by William Herschel more than two centuries earlier.

The main worry concerned the scan platform carrying the main camera. At first nobody was sure why it had given trouble, and there were fears that it might have been jammed by an icy particle in the Saturn ring area. Luckily this was not so; it was simply a lubrication problem, and by slewing the platform at reduced speed the cameras could still be aimed in the right direction. Otherwise it would have meant rotating the whole space-craft, which would have been far from easy.

For a long time after leaving Saturn, Voyager 2 remained in what is termed a cruise mode. Contact was good – the usual receiving stations had been joined by the large radio telescope at Tidbinbilla in Australia – and all seemed to be well. On the other hand, it had to be admitted that Voyager 2 was by now an old probe. It had been launched in 1977, so that by the time of the scheduled Uranus encounter it would have been in space for well over eight years.

As we assembled at the Jet Propulsion Laboratory, in December 1985, there was a feeling of excitement which seemed to me to be even greater than that for the Jupiter or Saturn encounters. After all, we had known a good deal about the two inner giants, even though some of our guesses had turned out to be wrong. We knew much less about Uranus, and everyone – scientists and non-scientists alike – was ready for major surprises.

Certainly Uranus was different from the other three giant planets. First, there was the extraordinary tilt of the axis, which, as we have noted, amounts to 98 degrees relative to the orbital plane. This is more than a right angle, so that technically Uranus spins in a retrograde sense. At the time of the Voyager encounter, the south pole of the planet was turned towards the Sun and the Earth, and was in the middle of its immensely long period of daylight. The space-craft would approach its target pole-on, and this alone would make the view very different from anything which had been seen at Jupiter or Saturn.

Impression of Voyager 2 passing Uranus.

Because virtually no detail can be seen from Earth, it was uncertain whether or not Uranus had any cloud-belts or bright zones, and it was thought that the disk might be almost blank even from close range, with the atmosphere clear to great depths. We also knew that Uranus had no strong source of internal heat, so that its effective temperature is much the same as that of Neptune even though Uranus is so much closer to the Sun.

In 1977 a system of rings had been discovered, not by direct observation but by what is termed the occultation method. Uranus had passed in front of a star, and hidden it. Both before and after the actual occultation the star had 'winked', so that obviously it was being briefly covered by a ring system round the planet. Subsequently it has been found that the rings are quite unlike those of Saturn; they are much thinner, and instead of being bright and icy they are as black as coal-dust. They had been confirmed by later occultations, and recorded by direct infra-red photography, but not much else was known about them. Finally there were the satellites, of which five were known: in order of distance from Uranus they had been named Miranda, Ariel, Umbriel, Titania and Oberon. All were smaller than our Moon, and all moved in the same plane as the Uranian equator. It was expected that they would be icy and cratered, but by now we were well aware that satellites of major planets have a habit of producing surprises, and we could only await events.

The first important discovery came on 30 December 1985, almost a month before closest approach. The Voyager pictures showed a new satellite, closer-in than any previously known members of the Uranian family even though it was still well clear of the outer ring. It was then that the N A S A planners again showed remarkable resource. They found that on 'encounter day', 24 January 1986, Voyager would pass within 323,000 miles (520,000 km) of the new satellite, and the chance of taking a picture was really too good to be missed. Hasty adjustments to the programme were made, and the satellite was duly photographed. It is small – only 96 miles (154 km) in diameter – and it has a darkish surface, with three obvious craters. Later, names were chosen; following the Shakespearian pattern, the satellite itself was christened Puck, while the craters were given the unusual names of Bogle, Lob and Butz.

This was only the beginning. As Voyager 2 drew in towards its target, further small bodies were discovered; one astronomer commented that God must have taken a shaker and scattered satellites in all directions. All were even smaller than Puck, and no details could be seen on them, but they were significant none the less. The inner two, Cordelia and Ophelia, move on opposite sides of the outermost ring, and act as shepherds, keeping the ring particles firmly in their orbits. Efforts to locate other shepherds, deep inside the main ring system, met with no success.

During the approach to Uranus there was a sudden diversion. A message from Europe indicated that there were problems in contacting the Giotto space-craft which was on its way to Halley's Comet: could the Voyager tracking system be 'borrowed' for a few hours? It says much for inter-national co-operation that the answer was an immediate 'Yes'. Fortunately the Halley problem proved to be minor, but it caused something of a sensation for a few hours.

Oberon, taken from 41,000 miles (660,000 km); the resolution is down to 7.4 miles (12 km). Note the mountain, which is 3.7 miles (6 km) high, projecting from the limb to the lower left.

Meanwhile, Uranus showed up as a pale, featureless disk. There was no sign of radio emission, and certainly there were no vivid colours of the type we had found on Jupiter or even Saturn. Then, a few days before closest encounter, the first clouds were seen. They were far from conspicuous, but they definitely existed, and they made it possible to measure the length of Uranus' axial rotation period – which proved to be 17 hours 14 minutes, rather longer than had been believed before the Voyager mission.

Daily Press conferences were held for the benefit of the media, but as the days went by, with little definite news, I heard comments that Uranus might prove to be a rather dull sort of world. Nothing could have been further from the truth. Suddenly, radio emissions were picked up, together with indications of a magnetic field. It was then found that the magnetic field was strange; it is reversed with respect to the Earth's, so that what we call the north pole of rotation is 'magnetic south', but in any case the magnetic axis is inclined to the rotational axis by almost 60 degrees. Moreover, the magnetic axis does not pass through the centre of the globe, but is considerably offset.

Here again Uranus differs from Jupiter or Saturn. It was suggested that it might be going through a 'magnetic reversal', or that the dynamo region responsible for the field was comparatively close to the planet's surface, but we did not pretend to know. The magnetosphere is quite extensive. It spreads out to 370,000 miles (600,000 km) on the day side of Uranus and 3,700,000 miles (6,000,000 km) on the night side, so that all the satellites, including the remote Oberon, are immersed in it. And ultra-violet observations showed strong emissions on the day side of the planet, producing

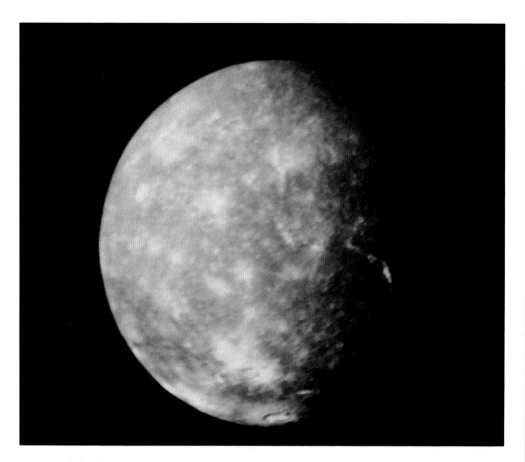

an 'electroglow' which again is unlike anything else we have so far found in the Solar System.

Several new rings were discovered, though apart from Cordelia and Ophelia there seemed to be no shepherd satellites. Before closest encounter it had looked as though the ring system might be more or less dust-free, but the last pictures, obtained when Voyager 2 had started on its outward journey, showed otherwise. I was in the main receiving room when they first came through, and for a moment I thought that I was looking at a view of Saturn's system; there was abundant dust, so that clearly there is a much greater quantity of thinly-spread material than had been anticipated.

Quite apart from all this, there were new ideas about the make-up of Uranus itself. The old model of a rocky core, surrounded by a liquid ocean of melted ices (ammonia, methane and water) overlaid by the deep, clear atmosphere, seemed to be in need of revision. There is certainly a rocky core, but it is now thought to be surrounded by a region in which the gases and ices are mixed in a dense layer, with the clouds above. The temperature at the visible surface is so low that methane is able to condense above the other clouds. Surprisingly, the dark pole was found to be two or three degrees warmer than the sunlit pole, presumably because it had not had time to cool down after its prolonged summer.

With the satellites, everything had to be carried through at breakneck speed. On 'encounter day', 24 January, all the pre-Voyager satellites were surveyed in rapid succession. They proved to be far from identical. The two largest, Titania and Oberon, are between 900 and 1,000 miles (1,400 and 1,600 km) in diameter; both showed craters, and both were icy. Oberon is

Titania, photographed from 310,750 miles (500,000 km); the resolution is down to 5.6 miles (9 km). Messina Chasmata is shown to the right of the disk. (Left)

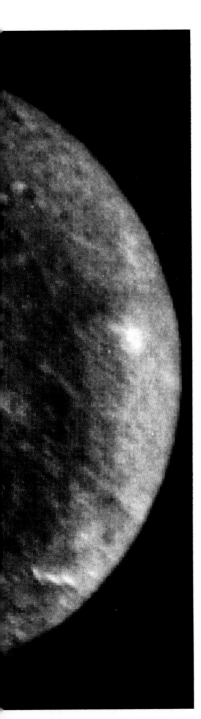

grey, and some of its main craters have dark material inside them, presumably due to a mixture of ice and carbon-rich material erupted from below. Near the crater now named Macbeth there is what seems to be a high mountain projecting from the limb, but we cannot be sure of its exact nature; remember, our maps of the satellites are very incomplete, because their opposite hemispheres were in darkness as Voyager passed by. Also we were seeing them pole-on, as with Uranus itself.

Titania, very slightly larger than Oberon, had distinct evidence of past activity upon its surface; there are ice-cliffs, faults, craters and valleys – one of which (Messina Chasmata) is at least 1,000 miles (1,600 km) long. Umbriel, smaller than Titania or Oberon, is much darker and more subdued, so that there must be a greater quantity of dark material mixed in with the ice. Could it be that some past event – perhaps a collision – has coated the surface with a more or less uniform, blackish layer? But there is one feature, Wunda, which is around 87 miles (140 km) across, and is decidedly bright. Unfortunately it lies right on the limb in the best Voyager picture, which means that it is almost on Umbriel's equator. (Since the pole lies in the centre of the visible disk, the equator must stretch all the way round the limb.)

Ariel is different again. It was found that as well as the expected craters, there were broad, branching, smooth-floored valleys which looked very much as though they had been cut by liquid. Ariel has no atmosphere, and is not massive enough to retain one, so that the valleys must have been formed by the expansion of the crust when the interior of the satellite froze solid, but undoubtedly they are very peculiar features.

Umbriel, taken from 346,000 miles (557,000 km); Wunda appears at the top of the picture – remember that this is on the satellite's equator. (Above)

Ariel, photographed from 106,000 miles (170,000 km). The resolution is down to 1.9 miles (3 km). Note the numerous valleys and fault scarps. (Right)

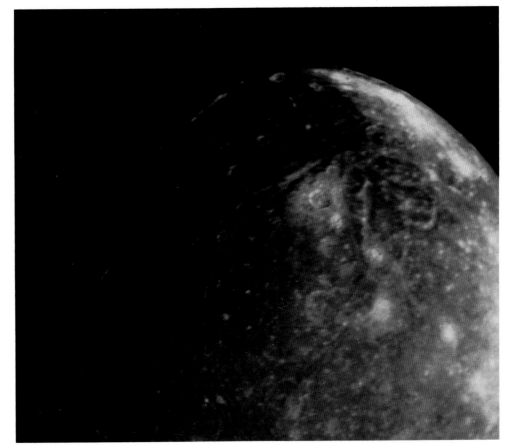

Miranda, shown in a computer-assembled mosaic of images obtained on 24 January 1986 by Voyager 2. Nine images were used to combine this full-disk, south polar view.

However, all these satellites paled in interest compared with Miranda, which is very small (less than 300 miles/500 km in diameter) and had been discovered, by Gerard Kuiper, as recently as 1948. Voyager passed it at a distance of only 1,860 miles (3,000 km), and sent back pictures giving a resolution down to 2,000 feet (600 m), so that our views of it are actually as detailed as those of any other world except those upon which space-craft have actually landed. My comment, on first seeing the pictures as they came through, was: 'You name it, Miranda has it!'

The landscape is almost incredibly varied. There are old, cratered plains; brighter areas with 12-mile (20-km)-high ice-cliffs; scarps, valleys, and two large, roughly rectangular regions which were nicknamed the 'Race-Track' and the 'Chevron', though they have now been given the more prosaic names of Inverness Corona and Arden Corona. In another age, no doubt Arden Corona, with its weird aspect, would have been interpreted as being artificial. To most of us at the Jet Propulsion Laboratory, Miranda represented the highlight of the whole Uranian mission.

How can Miranda be explained? There are suggestions that early in its history it was hit by a large body and shattered, subsequently re-forming; but this would involve the production of intense heat, and the icy Miranda is so small that this seems unlikely. It will take planetary geologists a very long time to come to any definite conclusions, and we may have to wait for the pass of another space-craft, which, sadly, is not likely to happen for many years yet.

By the end of the encounter it was clear to all of us that Uranus, far from being dull, had more than come up to expectations. Once again Voyager 2 had performed perfectly. As it had neared its target, an on-board computer fault had actually been put right by a command sent from the Deep Space Network at Pasadena – a running repair carried out thousands of millions of miles away in space. To complete its triumph, Voyager made its closest approach to Uranus almost exactly on schedule. To be precise, it was 1 minute 9 seconds early.

We left the Jet Propulsion Laboratory well satisfied. And Neptune lay ahead.

Miranda, photographed from 22,400 miles (36,000 km); resolution 2,600 feet (800 m). A remarkable ice-cliff appears to the right of the picture.

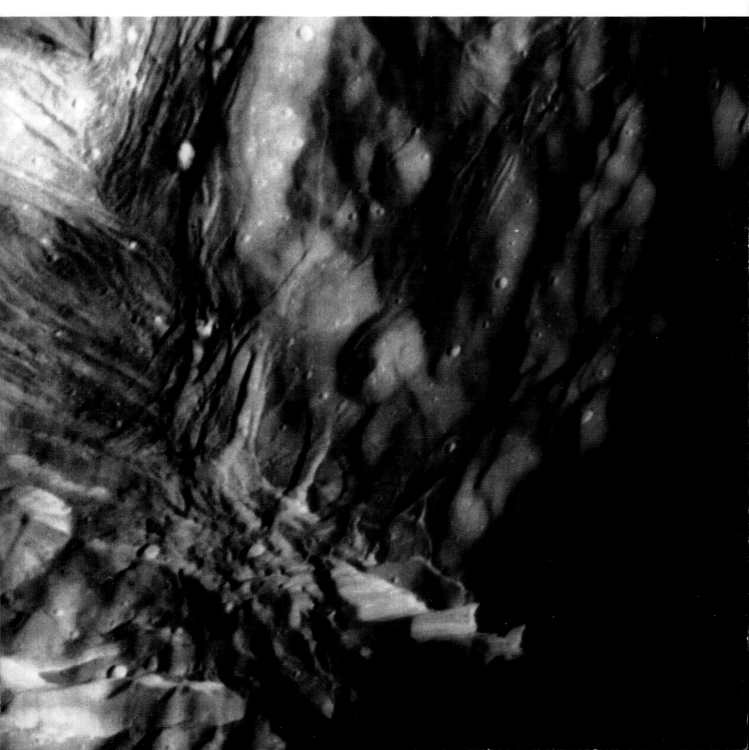

12

NEPTUNE: THE LAST TARGET (1989)

Neptune has never been regarded as a glamorous planet. It was discovered in 1846 as a result of mathematical calculations by two astronomers, one French and one English (an episode which nearly led to a full-scale international incident – but that is another story). It is much too far away to be seen with the naked eye, and it takes almost 165 years to go round the Sun.

In some ways it is very like Uranus; it is very slightly smaller, rather more massive, and with the same sort of surface. On the other hand it does not share Uranus' unusual axial tilt, and unlike Uranus it has considerable internal heat, so that it was thought likely to be a more dynamic world than its bland 'twin'.

The satellite system was known to be unlike any other. Before the Voyager mission only two moons had been discovered. Of these, one, Triton, was the only large planetary satellite to have retrograde motion – that is to say, motion in the direction opposite to that of the spin of its primary planet. Its distance from Neptune was a mere 219,000 miles (353,000 km), and the orbital period 5.88 days. Nobody was at all sure about its diameter; estimates ranged between 1,860 miles (3,000 km) and as much as 3,500 miles (5,600 km). It was thought that Triton might be larger than Mercury, and to have an atmosphere cloudy enough to hide its surface. Oceans of methane were suggested. On the contrary, Nereid, the other moon, was very small, and had a strangely eccentric orbit more like that of a comet than a satellite. According to one theory, Triton was not a genuine satellite of Neptune, but an interloper which had been captured in the remote past.

All in all, our knowledge of the Neptunian system was very meagre, and we realized that Voyager 2 carried all our hopes. If it failed, we would have to wait for many years before another probe could be dispatched – partly because of the lack of available money and partly because we could no longer use the 'gravity-assist' technique involving Jupiter, Saturn and Uranus one after the other. It was also painfully clear that Voyager 2 would be tested to its utmost limit. It was an old probe; after all, it had been planned, built and launched in the 1970s, and ever since 1978 it had been operating on its back-up receiver, so that there was always the danger of a final breakdown in communication. It says much for the NASA scientists that they were not only able to keep Voyager operating, but even to improve its performance by updating the Earth-based equipment.

Neptune is a long way away. At the time of the Voyager pass it would be precisely 2,750,876,750 miles (4,427,106,475 km) from Earth, so that a radio message from the space-craft would take 4 hours 6 minutes to reach us. By that time Voyager would have covered a grand total of over four and a half thousand million miles (seven thousand million kilometres) since being launched in 1977. The signal strength would be only 1/36th of that at the time of the Jupiter encounter, and new radio telescopes were brought in to help; in addition to the instruments at Goldstone in America, Robledo in Spain, and Parkes and Tidbinbilla in Australia, the twenty-seven 90-foot (28-m) antennæ of the V L A (Very Large Array) in New Mexico were involved, as well as a large radio telescope at Usuda in Japan. It was a truly international effort. Eventually, after a course correction on 1 August which altered Voyager's speed by 2.1 mph (3.3 kph), the probe reached its target point within six minutes of the scheduled time.

When I reached the Jet Propulsion Laboratory on 10 August 1989, I found an air of unsuppressed excitement. We all knew that this was not only the most potentially fascinating encounter of all, but it would also be the last for many years. Ideas of sending Voyager on to Pluto had had to be abandoned – it would have meant burrowing deep into Neptune's globe, which did not seem very practicable. Therefore, we were coming to the end of an era.

The Neptune encounter would be unlike any of the other Voyager passes, because everything would happen so quickly. At the JPL we were operating on Pacific Daylight Time, which is eight hours behind GMT. Voyager was scheduled to make its closest approach to Nereid at 9.13 p.m. on 24 August, sweep 3,000 miles (5,000 km) over Neptune's north pole at 1.02 a.m. on 25 August, and make its closest approach to Triton – 24,000 miles (38,000 km) – at 6.16 a.m. Therefore, the main picture-taking period would be compressed into less than 12 hours, and there could be no second chance.

Voyager 2. This is a full-scale model of the space-craft which by-passed all the four giant planets – Jupiter (1979), Saturn (1981), Uranus (1986) and Neptune (1989). I photographed the model at the Jet Propulsion Laboratory during the Neptune pass.

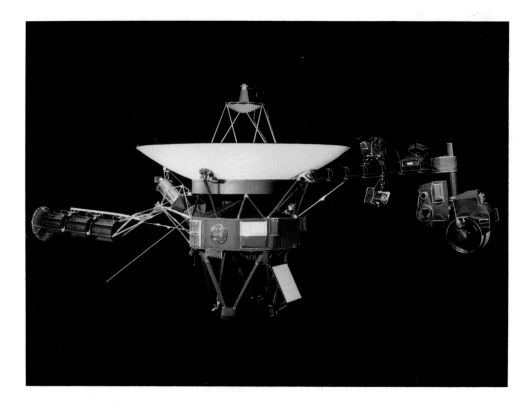

Before long it was clear that we were not going to be disappointed. First came the discovery of a system of relatively small inner satellites. One of them, now named Proteus, was actually larger than Nereid, and its diameter was estimated as 260 miles (420 km) but, unlike Nereid, it was too close to Neptune to be detected from Earth. Hasty adjustments to Voyager made close-range pictures of it possible, and it proved to be somewhat irregular in shape, with a dark surface and one large crater. Images were also obtained of the second of the new moons, Larissa, which also was darkish and rather irregular.

Nereid was the one 'miss'. It was in the wrong part of its very elliptical orbit, and Voyager could do no more than take a picture of it from a range of 2,890,000 miles (4,653,000 km), showing a darkish surface with a few light patches. It was a pity, but we had the satisfaction of knowing that on the outward journey we would – with luck – obtain splendid views of Triton, which was likely to be far more interesting.

Next came the rings. Before the mission, occultation techniques had indicated that there might be ring arcs – that is to say, incomplete rings – but we could not be sure, and the presence of Triton, a large, retrograde satellite, was expected to make the situation rather unstable. However, ring arcs were soon detected. 'Shepherd' satellites had been expected (as with Saturn's F-ring), but things turned out to be less straightforward. One of the arcs was associated with the fourth of the new satellites, Galatea,

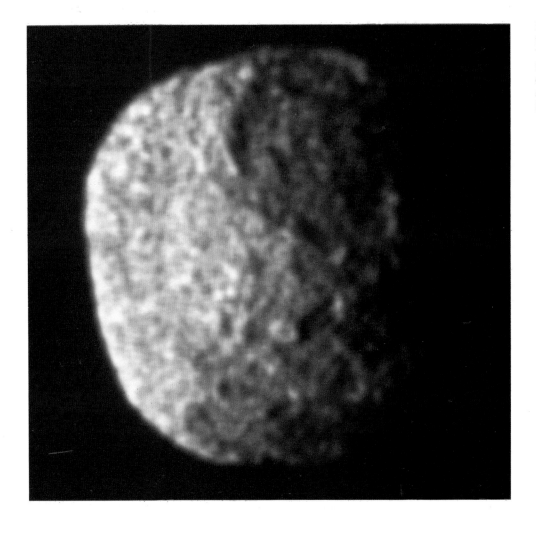

New satellite (Proteus); 25 August, from 91,000 miles (146,000 km). The apparent graininess of the image is caused by the short exposure necessary to avoid smearing. The satellite is dark and irregular in shape.

The rings of Neptune; Voyager 2 wide-angle camera, 26 August 1989, from 175,000 miles (280,000 km). The two main rings are seen; the bright arcs in the main ring were unfortunately on the opposite side of the planet for each exposure. Also shown is the faint inner ring at 26,000 miles (42,000 km) from the centre of Neptune, and the faint band or plateau which extends from the 33,000-mile (53,000-km) ring to roughly halfway between the two main rings. The bright glare in the centre is due to over-exposure of the crescent of Neptune.

and the other with the third of the newcomers, Despina, but the idea of a single 'shepherd' for even a partial ring was novel. Next it was found that the arcs were only the brightest portions of complete rings, and it was established that there are three rings altogether, plus a broad sheet of particles. The brightest part of the main ring contained bodies a few miles across which have been termed 'moonlets', though whether they are compacted bodies or mere clumps of particles is still unclear. The faintest sections of the rings were almost undetectable, and were only just above Voyager's threshold of visibility.

Efforts were made to judge the sizes of the dark ring-particles. If particles bounce light back in the general direction of the source (causing what is known as backscatter), they must be relatively large; if they scatter the light forward, they must be very small. It was found that the largest amount of fine material was contained in the 'plateau' which lay just outside the true inner ring; all the other rings contained a greater percentage of larger particles. (If you want to see just what is meant, try driving towards the setting sun in a car with a dirty windscreen. You will find it rather difficult, because the sunlight is forward scattered.) But even in the main Neptunian ring, the particles occupy about only 10 per cent of the total space, so that they are very flimsy compared with the rings of Saturn or even Uranus.

It had been assumed that Neptune must have a magnetic field, but though radio emissions were soon detected, there was a delay before Voyager passed through the bow shock – that is to say, the region where the solar wind is heated and deflected by interaction with Neptune's magnetosphere. The bow shock was finally recorded at a distance of 546,300 miles (879,000 km) from the planet. The magnetic field itself was weaker than those of the other giants but was, surprisingly, found to be inclined to the rotational axis by almost 50 degrees. Uranus, of course, has an even greater difference between its rotational and magnetic axes, but this had always been put down to the remarkable tilt of the rotational axis itself, and nobody had anticipated that from a magnetic point of view Neptune would be so like Uranus and so unlike Jupiter or Saturn. Moreover, the magnetic axis did not even pass through the centre of Neptune's globe, but was offset by a full 6,000 miles (10,000 km). Dr Norman Ness, principal investigator of the Voyager magnetic field experiment, commented that because of the tilt, and because of the offset, the dynamo electric currents responsible for the field must be close to the planet's surface – also that we would have to abandon the 'Act of God' theory that we had happened to catch Uranus just at the time when it was going through a magnetic reversal.

There had been speculation as to whether there would be any sign of Neptunian auroræ, the equivalent of our polar lights. In view of the presence of a magnetic field, auroræ seemed probable – and they were duly found; but instead of being near the geographical pole, they were closer to Neptune's equator. Remember, the rotational pole and the magnetic pole on the planet are a long way apart. At this time, of course, it was the northern hemisphere which was in the middle of its long winter.

As Voyager closed in, details on Neptune's beautiful blue surface became more and more striking. Pride of place went to a huge, Earth-sized feature which was instantly named the Great Dark Spot, in the southern hemisphere, which was likened to the Great Red Spot on Jupiter – though, in fact, the Great Dark Spot is not red; it is simply slightly less blue than its surroundings. Hanging above it were bright clouds which looked like cirrus, made up not of water ice but of methane ice; between the cirrus and the main cloud deck there was a clear area of around 30 miles (50 km). The cirrus was found to change quickly, while the Great Dark Spot itself has a rotation period slightly longer than that of its adjacent regions, so that it lags in longitude. Other features included a second dark spot, further south in latitude, plus a feature nicknamed the Scooter because of its relatively quick rotation; every few Neptunian days it 'catches up' the Great Dark Spot and laps it. The Scooter has a bright centre, but changes shape quite rapidly.

There are violent winds on Neptune; in some latitudes they blow retrograde at up to 700 mph (1,120 kph). The cirrus clouds above the Great Dark Spot do not take part. To quote Dr Bradford Smith, head of the Voyager imaging team, 'The clouds themselves aren't moving, but the atmosphere around them is. On Earth, lenticular clouds hang above a mountain without shifting, though the wind blowing around them and through them moves at high speeds.' Neptune was turning out to be a truly dynamic planet – a striking contrast to the more placid Uranus.

What about the atmosphere? Again the occultation procedure was used; the bright star Nunki (Sigma Sagittarii) was hidden by the planet, so that just before and just after the actual occultation its light came to Voyager by way of Neptune's atmosphere. Hydrogen is the main constituent (85 per cent) followed by helium (13 per cent). Methane, which is a hydrogen compound (CH_4), accounts for 2 per cent, but the clouds are largely of methane, and a definite cycle was established:

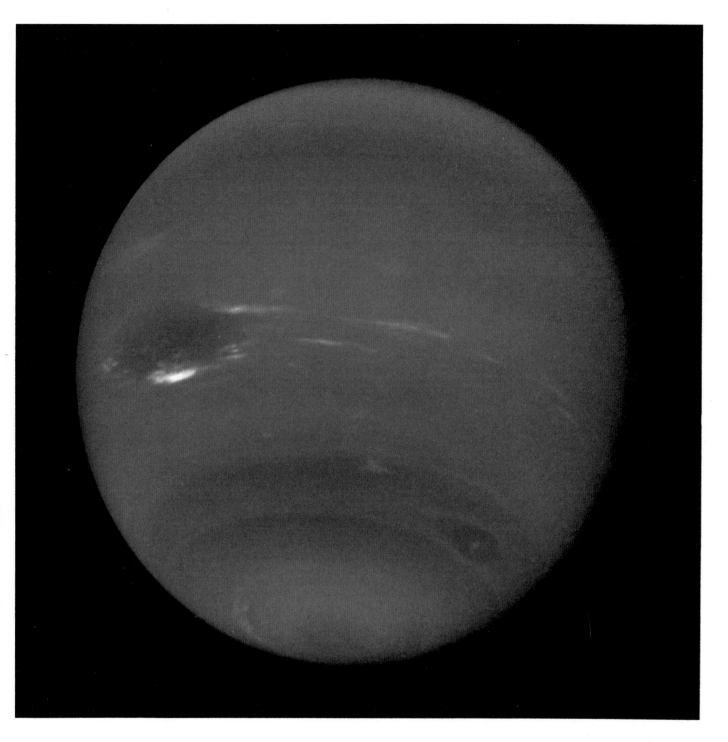

Neptune, 16–17 August 1989. The Great Dark Spot is dominant in this view, at latitude 22°S, and the smaller dark spot, at 54°S, can be seen near the terminator (lower right edge).

1. Solar ultra-violet radiation destroys methane high in Neptune's atmosphere by converting it to hydrocarbons, such as ethane and acetylene.

2. These hydrocarbons sink to the cold, lower stratosphere, where they evaporate and condense.

3. The hydrocarbon ice particles fall into the warmer troposphere, or lower atmosphere, where they evaporate and are converted back to methane.

4. Buoyant, convective methane clouds then rise up to the base of the stratosphere or higher, returning methane vapour to the stratosphere and preventing any net loss of methane.

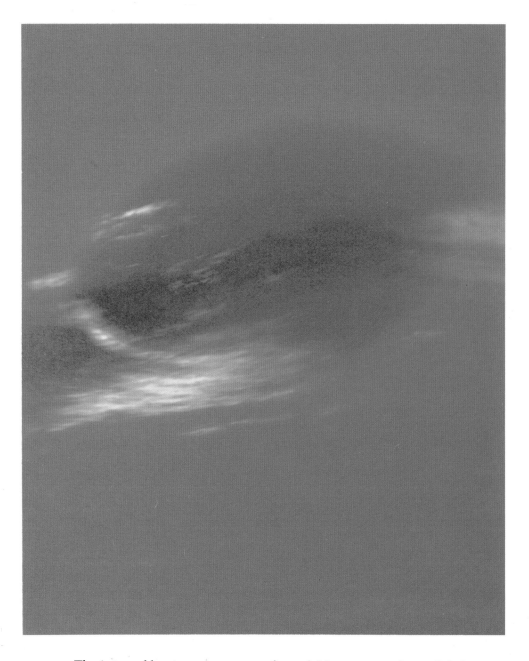

Neptune's Great Dark Spot; Voyager 2, 21 August 1989, from 1,740,000 miles (2,800,000 km); resolution down to 30 miles (50 km). The image shows the cirrus clouds overlying the boundary of the light and dark regions.

The internal heat-source was confirmed; Neptune sends out 2.8 times as much energy as it would do if it depended entirely upon what it receives from the Sun, as against no more than 0.1 per cent for Uranus. Deep below the clouds – say at 30 miles (50 km) or so – there may be other substances, including hydrogen sulphide.

As Voyager skimmed over Neptune's darkened north pole at 60,000 mph (96,000 kph) no pictures could be received; the speed was simply too fast, and any images would have been hopelessly smeared. The next event was the crossing of the ring-plane, and this was awaited with some apprehension, because nobody was sure whether or not the rings would be dense enough or contain enough particles to damage the space-craft. Just before the moment of crossing the number of impacts increased alarmingly. It was not long before the danger was over, but Norman Ness admitted that the situation had been 'scary'.

The clouds of Neptune. This Voyager 2 high-resolution colour image, taken two hours before closest approach, shows the cirrus clouds casting shadows on to the cloud-deck below. The width of the cloud streaks ranges between 30 to 125 miles (50 to 200 km) and their shadow widths range from 20 to 30 miles (30 to 50 km). The range is 97,500 miles (157,000 km).

Just over five hours later, Voyager encountered its very last target – Triton. We had half-expected this to be the highlight of the entire mission, and so it proved. Triton is an amazing world – in the words of Dr Laurence Soderblom, one of NASA's leading planetary geologists, 'the most mysterious thing we've ever seen'. We need not have feared that the surface would be hidden by clouds; Triton's atmosphere is very thin, with a ground pressure of only 0.01 of a millibar, and is made up of nitrogen, with a considerable amount of methane at lower levels. Haze was seen around the limb, but that was all, even though the tenuous mantle was found to extend upward for several hundreds of miles.

Triton was smaller than had been expected. Its diameter is only 1,681 miles (2,705 km), less than that of our Moon; this means that the density is considerable – more than twice that of water – and the globe must contain more rock and less ice than do the junior satellites of Saturn. The highly reflective surface is intensely cold, and indeed Triton, with its temperature of −400 °F (−236 °C) is the chilliest place ever visited by a spacecraft. But as Voyager 2 closed in, surprise followed surprise.

First, there is very little surface relief – no more than a few hundred feet – and high peaks and deep valleys are absent from Triton. Craters are scarce, and there are none more than around 17 miles (27 km) across, but there are extensive flows up to 50 miles (80 km) wide, due probably to ammonia-water fluids sent out from below the surface. The Neptune-facing hemisphere, imaged by Voyager, is divided into three regions: Bubembe Regio (western equatorial), Monad Regio (eastern equatorial), and Uhlanga Region (polar) – again, these names have been given by the nomenclature committee of the International Astronomical Union. It is Uhlanga which is so striking, because it is covered with pink snow – not ordinary snow of the same type as ours, but nitrogen snow; Triton is so cold that nitrogen freezes. In fact there must be an underlying layer of water ice, because other ices are not hard enough to maintain any kind of surface relief over long periods, but the pinkness of the polar cap gives it a truly weird appearance.

This is not all; there are active geysers. At a depth of from around 60 to 100 feet (20 to 30 m) the pressure is bound to be high, so that nitrogen can exist as a liquid. If this liquid nitrogen can migrate upward, it will come to a region where the pressure is around one-tenth that of the Earth's air at sea-level. In this case, said Dr Laurence Soderblom, one of NASA's leading experts, the nitrogen will explode into a shower of ice and gas – roughly 80 per cent ice and 20 per cent vapour – and will travel up the nozzle of the geyser-like vent at up to 50 feet per second (150 m per second), fast

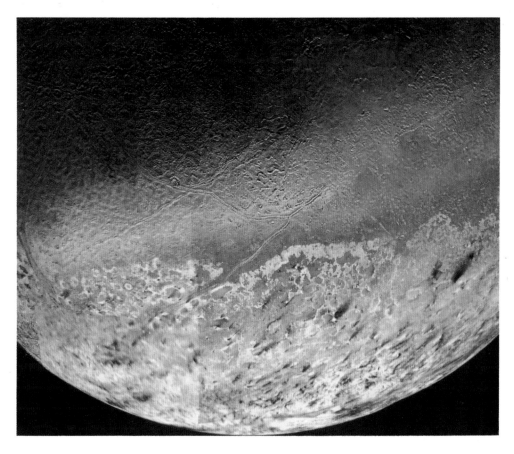

Triton, 29 August 1989; Voyager 2. A dozen individual images were combined to produce this view of the Neptune-facing hemisphere. The large south polar cap is pinkish; it may consist of a shrinking layer of nitrogen ice deposited during the previous winter. North of the ragged edge of the cap the satellite is darker and redder, perhaps because of the action of ultra-violet radiation upon methane ice. Running across this darker region, roughly parallel to the edge of the cap, is a slightly bluish layer.

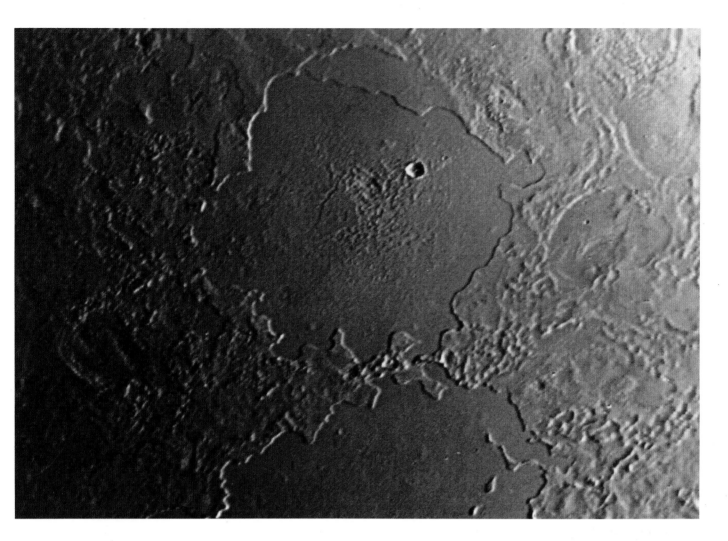

Frozen lake on Triton; 25 August 1989. The view is about 300 miles (500 km) across. The resolution is about 3,000 feet (900 m).

enough to make it rise to a height of around 30 miles (50 km) before falling back. The outrush sweeps dark débris along, and this débris is wafted downwind to produce the dark plumes which were evident in the Voyager pictures. Certainly nothing of this sort had been expected, but it seems clear that Triton joins the Earth, Io and Venus as being known worlds which are actively volcanic, even though Triton's geysers are very different from the volcanoes of other active worlds.

North of the pink cap there is a darker, redder transition region, and running across this is a slightly bluish layer which may be due to tiny crystals of methane scattering the incoming light. Monad Regio, further from the pole, shows shallow pits, mushroom-like features of uncertain origin, and 'frozen lakes' with flat, smooth floors; Bubembe Regio is characterized by what is called cantaloupe terrain, from its superficial resemblance to a melon-skin, crossed by huge, shallow fissures which meet in X and Y junctions.

Over a period of many years the whole aspect of Triton may change as the seasons cause alterations in temperature, but unfortunately not even the Hubble telescope can show surface detail, and we must wait patiently for a new probe. But everything indicates that Triton was once an independent body, which was captured by Neptune long ago. Originally its path round Neptune would have been eccentric, but over perhaps a

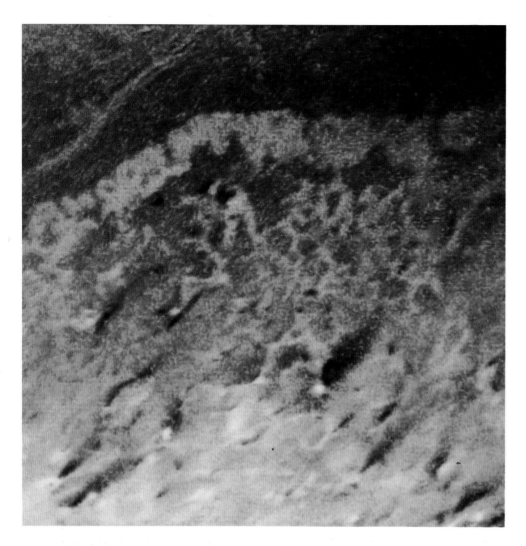

False-colour image of Triton; Voyager 2, 27 August 1989, from 118,000 miles (190,000 km). The smallest visible features are 2.5 miles (4 km) across. The image shows a geological boundary between dark and patchy materials; a layer of pinkish material stretching across the centre must be thin, as underlying albedo patterns show through. The dark streak seems to be one of Triton's ice geysers.

thousand million years the path was forced into the present circular form; during this time there would have been internal flexing and heating, coupled with surface activity. For example, mixtures of water and ammonia would have flowed out, filling the shallow regions which we see today as frozen lakes.

Within a few days the spectacle was over; by August 29 Voyager was already more than 4,500,000 miles (7,000,000 km) beyond Neptune, and Triton was no more than a speck in the distance. The farewell party at JPL was a mixture of elation, because of the success of the mission, and sadness, because we knew that it was the last time for many years that we would meet up in such a way – and for most of us (including me) the last time of all. But it had been a week never to be forgotten, and Laurence Soderblom summed it up very appropriately: 'Wow! What a way to leave the Solar System!'

13

RENDEZVOUS WITH HALLEY'S COMET (1986)

Though this book has been concerned with missions to the planets, I hope that I may be allowed to make what seems to be something of a digression, and say a little about the cosmic armada which went to Halley's Comet in March 1986. For one thing, there is an immediate link with the planetary programmes, because the two Russian space-craft were modelled upon the Veneras, and actually dropped balloons into Venus' atmosphere as they passed by that world. Secondly, comets are associated with meteors, and possibly even with some types of asteroids. According to one theory, the tiny 'earth-grazing' asteroids which occasionally fly past us at close range are nothing more than comets which have lost all their dust and gas, and have been left as small, inert lumps of icy material. And thirdly, the probes to Halley's Comet would never have been possible but for the experience gained in the earlier missions to the planets.

I have already said something about comets. All the short-period members of the swarm are faint, because they have been back to the Sun so many times that they have lost most of their volatiles. Halley's Comet is the one exception. During most of its historical returns it has been striking, and this was certainly the case in 1910, when it had a brilliant head or coma and a long tail. However, from the outset it was known that the 1986 return would be about as unfavourable as possible, because the comet would be on the far side of the Sun at the time of perihelion, and would be out of view from Earth. From Venus, things would have been much better. Of course, nobody on the surface of Venus could hope to see the comet or anything else, but at the time of Halley's perihelion it was hoped to use the orbiting Venus probe to obtain some data. Alas for our hopes – a violent solar storm cut off communication altogether during the vital period! Mother Nature was at her most un-cooperative.

On the other hand, there was no reason to doubt that space-craft could be dispatched, and no less than five missions were planned: two by the Japanese, two by the Russians, and one by the European Space Agency (ESA).

The main absentee from this programme was the USA. Their space planners were only too anxious to take part, but the financial controllers were not, and the ambitious American plan had to be abandoned, a fact which will no doubt be regretted for the next seventy years or so. However, there was one American experiment which was associated with cometary research. Way back in 1978 an earth satellite, International Sun-Earth Explorer 3 (ISEE3) had been launched to study the interactions between the solar wind and the Earth's magnetosphere. It performed well, and calculations showed that its orbit could be modified to take it to a rendezvous with a comet – not Halley's, but a much fainter object, Comet Giacobini-Zinner, which has a period of six and a half years, and which had been observed regularly since it had originally been found in 1900. Estimates of the coma diameter gave a value of 30,000 miles (50,000 km); the tail was expected to be up to 300,000 miles (500,000 km) long, and the comet was associated with a yearly meteor shower which gave occasional brilliant displays, as had happened in 1933 and again in 1946. Giacobini-Zinner was better than nothing at all, and at least there would not be the cost of building and launching a new space-craft. A series of complicated manoeuvres followed, including five close passes of the Moon (on 22 December 1983 ISEE3, now re-named ICE or the International Comet Explorer, was only 75 miles/ 120 km above the lunar surface), and finally, on 11 September 1985, the probe passed through the comet's tail, moving at a relative speed of 13 miles (21 km) per second. It carried no cameras, but it did send back very useful information about the particle density and the electrical and magnetic conditions near the comet, so that it must be regarded as a success. Later, on 25 March 1986, it passed by Halley's Comet at around 19,000,000 miles (30,000,000 km), though by then the results from the main armada had been obtained, and it cannot be said that ICE had much to contribute. It is still under control; it should pass close to the Moon again on 10 August 2014. One American astronomer has suggested that it ought to be fished down and given a place of honour in the Smithsonian Institute!

Halley's Comet had been lost to view since 1911, but in 1984 astronomers at the Palomar Observatory recovered it as a tiny smudge of light. As it neared the Sun it began to brighten up; from my observatory I had my first view of it on 11 September 1985, though I could make out no tail. Meanwhile, the various space-craft were on their way. The Russian vehicles were dispatched first, Vega 1 on 15 December 1984 and Vega 2 six days later. (Please note the name 'Vega' has nothing to do with the bright star Vega or Alpha Lyræ, which, as seen from Britain, is almost overhead during summer evenings.) The Japanese probes followed in 1985, Sakigake on 8 January and Suisei on 18 August. Both these were very small, and cylindrical in shape; their main rôles were to fly past the comet at a respectful distance, studying the environment and making measurements of particles and various electrical and magnetic fields. This was the first time that the Japanese had attempted to send any vehicles beyond Earth orbit. Knowing their efficiency, I expected them to succeed, and they did.

On 2 July 1985 I was at the French base at Kourou, in Guyana, watching the launch of the European probe, which had been named Giotto in honour of the Italian artist who had painted a famous picture of the comet in 1604. The launch vehicle was an Ariane rocket. I cannot say that I was entirely

Halley's Comet on 9 December 1985, taken through the 39-inch (1-m) Schmidt telescope, European Southern Observatory (La Silla, Chile).

confident, because the Ariane record was not impeccable, and there had been occasions when valuable payloads had been dumped in the sea; there was only one Giotto, and if Ariane failed us we would have to wait 75 years for another chance. The countdown, briefly halted for technical reasons, was tense by any standards. Then, through the window of my commentary box, I saw Ariane rise slowly and majestically from its launching pad. Gradually it picked up speed, and before long it had been lost to sight. After a few minutes, we learned that all was well – and I think that the cheer that went up must have been audible in Devil's Island, some way out across the sea. Giotto was on its way to Halley's Comet; its journey would take over eight months.

The space-craft was cylindrical in shape, with a diameter of 6.2 feet (1.9 m) and an overall height of 9.5 feet (2.9 m); at launch it had a total weight of 2,112 pounds (960 kg). It carried all manner of experiments, notably the HMC or Halley Multi-Colour Camera; there were bumper shields to protect it against the impacts of dust and ice particles, but nobody was sure how effective they would be. One problem was caused by the fact that Halley's Comet moves round the Sun in a wrong-way or retrograde direction, so that Giotto would have to meet it head-on at a relative velocity of over 42 miles (68 km) per second – and at this speed, a dust-particle weighing only a tenth of a gramme can penetrate an aluminium shield 3 inches (8 cm) thick. Giotto's bumper shield, designed by Fred Whipple, ιwas made up of two parts, so that the main impact would be absorbed by the outer skin and the inner layer would do the rest.

Though comets had been observed so often, not a great deal was known about them. The official theory, due to Fred Whipple, stated that the main mass was concentrated in a 'dirty ice' nucleus only a few miles across; according to a rival theory, championed by R. A. Lyttleton, a comet resembled a flying gravel-bank, with no definite nucleus at all. Yet another idea was supported by Sir Fred Hoyle and his colleague Chandra Wickramasinghe, who believed comets to be coated with dark organic material, and to be capable of carrying life from one part of the universe to another. They even went so far as to suggest that life was originally brought to Earth by way of a comet, and that there might even be diseases spread by comets which dumped bacteria into the Earth's upper atmosphere. To Hoyle and Wickramasinghe, comets are not true members of the Solar System, but are interlopers from outer space. This is still a minority view, but it is worth noting that much later (in 1989) it was supported by two eminent astronomers, J. Jones in Canada and Susan Wyckhoff in Arizona, who made measurements of the carbon compounds of Halley's Comet and concluded that an interstellar origin was very much on the cards.

I was a member of the International Halley Watch; my rôle was to photograph the comet as often as I could, using a small scale so that the pictures could be used for positional measurements. I was also involved in the BBC television commentaries from the ESA headquarters at Darmstadt, in West Germany. I arrived there a week or so before the scheduled Giotto encounter on the night of 13–14 March. Chandra Wickramasinghe telephoned me, urging me to broadcast his prediction that the comet's nucleus would be dark instead of ice-bright, as most people expected. I did as he asked, though frankly I did not expect him to be right.

In Darmstadt there were representatives from all nations, and there was absolute international collaboration, with Iron Curtains and even Bamboo Curtains conspicuous only by their absence. On 6 March Russia's Vega 1 opened the encounter by by-passing the comet's nucleus at 5,525 miles (8,890 km), and sending back pictures and data. Next came Japan's Suisei, at 4,350,000 miles (7,000,000 km) on 8 March, followed by Vega 2 at 5,000 miles (8,030 km) on 9 March and then Sakigake at 94,000 miles (151,000 km) on 11 March. All the data were promptly passed on, and final adjustments were made to the trajectory of Giotto, which was scheduled to penetrate the comet's coma and send back close-range pictures of the nucleus. Up to that time nobody had ever had a clear view of a cometary nucleus, which is always hidden from us by the dust and gas of the surrounding coma.

When our television programme opened, late in the evening of 13 March, nobody knew what was in store; whether Giotto would survive the battering it was bound to receive during its encounter with Halley was anybody's guess, but when the pictures started to come through it was clear that the camera was working well. At three minutes past midnight on 14 March, Giotto made its pass of the comet's nucleus at 370 miles (596 km). Unfortunately it had been hit, 7.6 seconds earlier, by a particle which was probably about the size and mass of a grain of rice, and Giotto had been jolted out of alignment, so that for the next half-hour the signals were intermittent. The

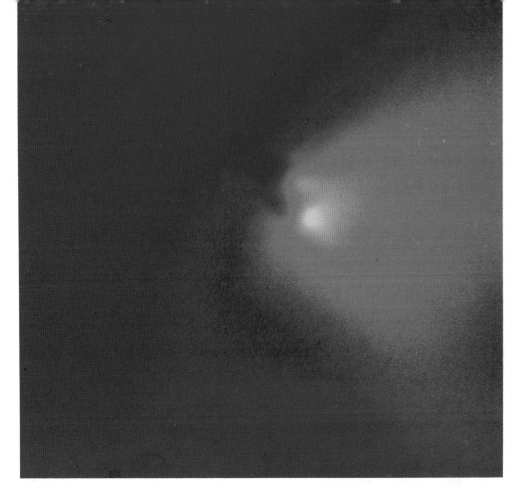

camera did not work again, and the last really good picture was that which had been sent back from a range of 1,200 miles (1,930 km).

The view was striking. As Hoyle and Wickramasinghe had predicted, the nucleus was indeed dark. Dr H. U. Keller, head of the Giotto camera team, described it as 'blacker than velvet', so that Fred Whipple's dirty ice-ball was much dirtier than had been anticipated.

We now know that the nucleus of Halley's Comet is rather irregular in shape, measuring 9.3 by 5 by 5 miles (15 by 8 by 8 km), with a total mass of from 50,000 million to 100,000 million tons – so that it would take 60,000 million Halleys to make up one body as massive as the Earth. Dust-jets were active from a small area of the nucleus on the sunward side, and the temperature was found to be 117°F (47°C), much higher than expected; it was inferred that the warm, dark dust covered an icy interior. There were also hills and craters. It seems that the nucleus is eroded by about 0.4 inch (1 cm) per day when the comet is near the Sun, so that at each return it loses at least 300,000,000 tons of material. It cannot last for ever – compared with planets, comets are short-lived things – but it is in no imminent danger of fading away, as some other periodical comets have been known to do.

I had a particularly interesting view of Halley's Comet in January 1989, with the 60-inch (1.5-m) Danish telescope at the La Silla Observatory in Chile; there was still a trace of the coma, but within a few years it will have passed beyond range, not to be back once more before the next perihelion passage in the year 2061. It is not outrageous to suggest that it will then be visited by a manned space-craft. In any case, the 1986 probes showed the

Nucleus of Halley's Comet, from 11,000 miles (18,000 km), taken from the HMC of Giotto. Again, two bright jets directed at the Sun are shown. The frame size of the image is 19 miles (30 km).

way, and in a few days they told us more about these strange, ethereal wanderers than we had been able to find out throughout the whole of human history.

By now Halley's Comet has returned to the depths of the Solar System, though in February 1991 it produced yet another surprise by a sudden outburst – brightening it by several magnitudes; the cause of this is still unclear. I had my last view of it in 1992 with the 60-inch (1.5 m) Danish telescope at La Silla, by which time the outburst had subsided and the comet had become totally inert – as it will remain until it draws inward once more in seventy years' time. But what of Giotto?

The probe had been a triumph; but its work was not over. After the Halley encounter it was put into hibernation, but in February 1990 it was reactivated, and it was sent on to an encounter with a faint periodical comet, Grigg-Skjellerup, on 10 July 1992. Many of the instruments were in excellent order (sadly, apart from the camera) and valuable results were obtained; Grigg-Skjellerup is a much smaller, older comet than Halley. Giotto suffered no major hits, even though it may have passed within 200 miles (320 km) of the comet's nucleus.

Unfortunately there is not enough fuel reserve to send the probe on to yet a third comet, but we know just where it is, and so to some extent we can still control it. Perhaps, one day, we may even be able to bring it home – frankly, I hope so!

14

INTO THE FUTURE

By the end of 1989, little more than three decades after Luna 3's flight round the Moon, all the planets in the Solar System had been by-passed with the exception of Pluto. Earlier plans to send a space-craft to this strange, gloomy little world had been ruled out by financial cutbacks, and there is no hope of a mission there yet awhile.

However, Nature has for once been helpful. In 1977 it was discovered that Pluto is not alone; it is accompanied by another body, Charon, which has more than half the diameter of Pluto itself. The two move in an unusual way. Pluto's rotation period is 6 days 9 hours, and this is also the revolution period of Charon, so that the two are 'locked' much as the two bells of a dumbell if you twist them by their connecting bar. During the 1980s the orbits were tilted to us at an angle which meant that Pluto and Charon periodically eclipsed or occulted each other. This made it possible to analyse their light individually, and even to draw up crude maps of Pluto's surface – a remarkable feat in view of the fact that the apparent disk of the planet is so small that its diameter is difficult to measure. Spectroscopic results indicate that Pluto has a surface coated with methane frost, and that it has an extensive though thin methane atmosphere. Charon has a surface layer of water frost, with no atmosphere at all.

Pluto cannot be classed as a bona-fide planet. It is too small, and it has the wrong sort of orbit. It and Charon may merely be the largest or closest members of a whole swarm of bodies moving beyond the path of Neptune. Some years ago Gerard Kuiper suggested that such a belt might exist, and recently we have found several remote bodies, all around 150 to 200 miles (250 to 320 km) across at most, which may well be Kuiper Belt objects.

There may even be a link here with another curious body, discovered by Charles Kowal from Palomar Observatory in 1977 and named Chiron. (The name is in honour of the kindly centaur of the Jason legend; it is a pity that it is so like Charon.)

Chiron has a revolution period of 50 years, and spends almost all its time between the orbits of Saturn and Uranus. Its discovery was unexpected, and certainly no asteroids were thought to exist in this part of the Solar System, but what else could Chiron be? It was given an asteroidal number, 2060; apparently it was fairly large, with a diameter of at least 125 miles (200 km). Spectroscopic results indicated that its surface was darkish, and possibly rocky.

Then, in 1987, it was found that Chiron was becoming measurably brighter than it ought to be, and in 1989 two American astronomers, K. J.

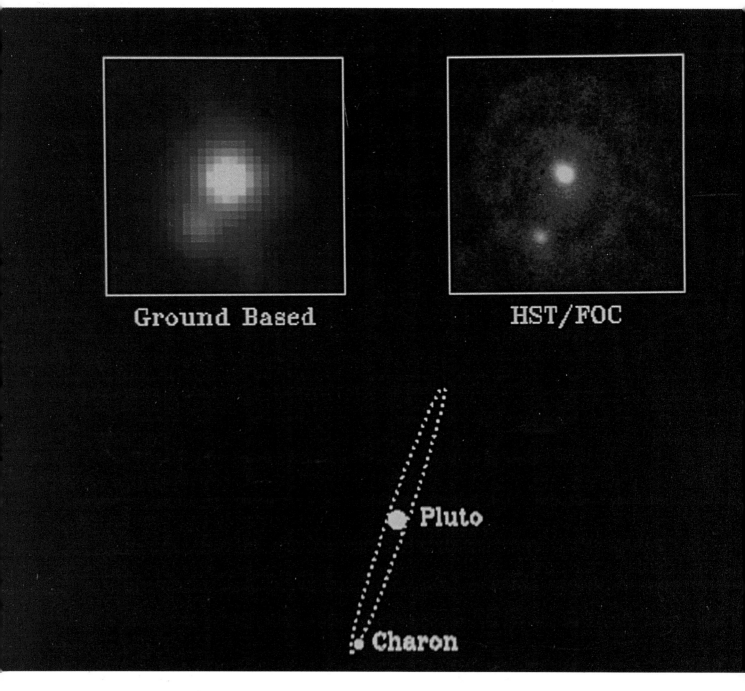

Ground Based

HST/FOC

Pluto

Charon

Meech and M. J. S. Belton, used electronic equipment with the 158-inch (4-m) reflector at Kitt Peak, in Arizona, to show that Chiron was developing a coma. The coma extended out to 5 seconds of arc, while the apparent diameter of Chiron itself is less than 1 second of arc. Could Chiron really be an immense comet rather than an asteroid, so that its ices are starting to evaporate as it draws inwards? Perihelion is due in 1996, when the distance from the Sun will be only 794,300,000 miles (1,278,000,000 km), less than the minimum distance of Saturn (837,000,000 miles/1,347,000,000 km). As yet we do not know, but certainly the nature of Chiron is very much in doubt, though it would be very dangerous to infer that Pluto and Charon could also be cometary.

Pluto and Charon; Hubble Space Telescope view compared with the best ground-based view. The relative orbit of Charon is shown above.

The best images of Pluto and Charon have been obtained with the Hubble Space Telescope, but of course no surface details can be seen, and we must wait for a new probe – which could be launched early in the twenty-first century.

It is possible that another large planet exists, far out beyond the orbits of Neptune and Pluto. Pluto was found not too far from the position predicted for it on the basis of slight irregularities in the movements of Neptune and Uranus; yet, as we have seen, Pluto is so small and so lightweight that it could not possibly drag either giant out of position by a measurable amount. Either the accuracy of the prediction was sheer luck, or else the real 'Planet X' remains to be located.

Various attempts have been made to work out where it may be. The trouble is that the wanderings of Uranus and Neptune are so slight that they are easily swamped by unavoidable errors of observation – and unless we have at least a fair idea of the position of Planet X, searching for it is much more futile than looking for the proverbial needle in the equally proverbial haystack. Moreover, one of the most promising investigations, carried out in America by Dr John Anderson, suggests that Planet X has a very eccentric orbit, so that at the moment it is out of range.

However, there is one possibility. The two outer-planet Pioneers and the two Voyagers are now leaving the Solar System permanently. Suppose that one of them passed close enough to Planet X to be measurably perturbed? This might give us a clue as to the whereabouts of Planet X itself. If all goes well, we should be able to keep in touch with all four space-craft for some years yet, when they will be well beyond the edge of the known planetary system. True, the chances of locating Planet X in this way are very slight, but they are not nil.

Eventually, of course, the Pioneers and the Voyagers will recede so far that we will lose all track of them. They will go on travelling between the stars for an indefinite period, until either they are destroyed by collision with a solid body or else collected by an alien civilization many light-years from us. Thoughtfully, they have been provided with plaques, pictures and recordings which would, it is hoped, give any 'finders' a clue as to their system of origin. The Voyagers even carry recordings of familiar Earth sounds, including the crying of a baby, a speech from the former Secretary-General of the United Nations (Dr Kurt Waldheim), and a performance from a pop group. We must assume that our alien friends are familiar with record-players, but it may well be that the sounds of the pop group will alone indicate that the Earth is best left firmly in isolation!

Finally, it is worth looking at some of the plans which are now under serious consideration – bearing in mind that politics and finance are all-important, and that already some particularly valuable projects have had to be cancelled because of the lack of funding. The collapse of the USSR has also caused problems; nobody really knows what is going to happen to the Russian space programme, and at the moment it is in a state of total uncertainty.

Some projects seem to be fairly safe. One of these is Cassini, due to be launched in 1996; it will reach its target – Saturn – in 2004, and, if all goes well, drop a small probe, Huygens, on to the surface of Titan, so that at last we ought to find out what lies below those tantalizing clouds. Much nearer home, Mars is sure to be a target; here there is close collaboration between the Russian and American programmes, and we may hope for a Mars rover as well as a sample and return probe.

Otherwise, we can only wait and see. No doubt the Shuttle will continue to be used, though we cannot be so certain about the Russian equivalent, Buran. Plans for a major space-station are well advanced, though here too everything depends upon how the world political situation develops. When a serious attempt to set up a fully-fledged Lunar Base will be made is equally uncertain; it could be within the next decade, or it may be postponed for a long time – and the same is true of the first manned expedition to Mars. But at least the opportunities are there. The first phase of planetary exploration has been dramatic; I have a feeling that the next phase, when it starts, will be more dramatic still.

Planet	Distance from Sun, millions of miles:			Distance from Sun, millions of km:			Revolution period	Rotation period (equatorial)
	max.	min.	mean	max.	min.	mean		
Mercury	43	29	36	69.7	45.9	57.9	87.97 days	58.65 days
Venus	67.6	66.7	67.2	109	107.4	108.2	224.7 days	243.16 days
Earth	94.6	91.4	93	152	147	149.6	365.26 days	23 hr 56 min 4 sec
Mars	154.5	128.5	141.5	249	207	227.9	687 days	24 hr 37 min 23 sec
Jupiter	506.8	459.8	483.3	816	741	778.3	11.86 years	9 hr 50 min 30 sec
Saturn	937	834.6	886.1	1,507	1,347	1,427	29.46 years	10 hr 39 min
Uranus	1,867	1,699	1,783	3,004	2,735	2,869.6	84.01 years	17 hr 14 min
Neptune	2,817	2,769	2,793	4,537	4,456	4,496.7	164.8 years	16 hr 3 min
Pluto	4,583	2,766	3,666	7,375	4,425	5,900	247.7 years	6 days 9 hr

Planet	Mean orbital velocity per second		Inclination of axis (degrees)	Orbital eccentricity	Orbital inclination	Equatorial diameter		Escape velocity per second	
	miles	km				miles	km	miles	km
Mercury	29.7	47.9	0	0.206	7° 00' 16"	3,033	4,879	2.6	4.3
Venus	21.8	35.0	178	0.007	3° 23' 40"	7,523	12,104	6.4	10.4
Earth	18.5	29.8	23.44	0.017	0° – –	7,928	12,756	6.94	11.2
Mars	15	24.1	23.98	0.093	1° 50' 59"	4,222	6,794	3.2	5.0
Jupiter	8.1	13.1	3	0.048	1° 18' 16"	89,424	143,884	37.1	60.2
Saturn	6	9.6	26.7	0.056	2° 29' 22"	74,914	120,536	22	32.3
Uranus	4.2	6.8	98	0.047	0° 46' 28"	31,770	51,118	13.9	22.5
Neptune	3.4	5.4	29	0.009	1° 45' 20"	31,410	50,538	15.1	23.9
Pluto	2.9	4.7	122.5	0.248	17° 12' –	1,444	2,324	0.7	1.2

Planet	Mean surface temperature		Mass (Earth = 1)	Volume (Earth = 1)	Density (water = 1)	Surface gravity (Earth = 1)	Maximum apparent diameter as seen from Earth (seconds of arc)	Mean diameter of Sun, as seen from planet
	°F day	°C day						
Mercury	+662	+350	0.055	0.056	5.5	0.38	12".9	1° 22' 40"
Venus	+896	+480	0.815	0.86	5.3	0.9	65".2	44' 15"
Earth	+72	+22	1	1	5.5	1	–	32' 01"
Mars	−9	−23	0.107	0.15	3.9	0.38	25".7	21' 00"
Jupiter	−238	−150	318	1,319	1.3	2.64	50".1	6' 09"
Saturn	−292	−180	95	744	0.7	1.16	20".9	3' 22"
Uranus	−353	−214	15	67	1.3	1.17	3".7	1' 41"
Neptune	−364	−220	17	57	1.5	1.2	2".2	1' 04"
Pluto	−382	−230	0.002	<0.01	2.2	0.06	<0".25	49"

APPENDIX II
SATELLITE DATA

Satellite	Mean distance from centre of planet		Orbital period (days)	Orbital eccentricity	Orbital inclination to planet's equator (degrees)
	miles	km			
Mars					
Phobos	5,800	9,270	0.3189	0.021	1.1
Deimos	14,600	23,400	1.2624	0.003	1.8
Jupiter					
Metis	79,530	127,960	0.295	0	0
Adrastea	80,160	128,980	0.298	0	0
Amalthea	112,680	181,300	0.498	0.003	0.45
Thebe	137,900	221,900	0.675	0.013	0.9
Io	262,030	421,600	1.769	0.004	0.04
Europa	416,970	670,900	3.551	0.009	0.47
Ganymede	665,000	1,070,000	7.155	0.002	0.21
Callisto	1,168,400	1,880,000	16.689	0.007	0.51
Leda	6,805,000	11,094,000	238.7	0.148	26.1
Himalia	7,134,900	11,480,000	250.6	0.158	27.6
Lysithea	7,284,000	11,720,000	259.2	0.107	29.0
Elara	7,295,000	11,737,000	259.7	0.207	24.8
Ananke	13,176,000	21,200,000	631	0.17	147
Carme	14,046,000	22,600,000	692	0.21	164
Pasiphaë	14,605,000	23,500,000	735	0.38	145
Sinope	14,730,000	23,700,000	758	0.28	153
Saturn					
Pan	83,033	133,600	0.57	0	0
Atlas	85,560	137,670	0.602	0.002	0.3
Atlas	85,560	137,670	0.602	0.002	0.3
Prometheus	86,600	139,350	0.613	0.004	0.0
Pandora	88,100	141,700	0.629	0.004	0.1
Epimetheus	94,110	151,420	0.694	0.009	0.3
Janus	94,140	151,470	0.695	0.007	0.1
Mimas	115,300	185,540	0.942	0.020	1.52
Enceladus	147,940	238,040	1.370	0.004	0.02
Tethys	183,140	294,670	1.888	0	1.86
Telesto	183,140	294,670	1.888	0	2
Calypso	183,140	294,670	1.888	0	2
Dione	234,570	377,420	2.737	0.002	0.02
Helene	234,570	377,420	2.737	0.005	0.2
Rhea	327,560	527,040	4.518	0.001	0.35
Titan	759,390	1,221,860	15.945	0.029	0.33
Hyperion	920,510	1,481,100	21.277	0.104	0.43
Iapetus	2,213,360	3,561,300	79.331	0.028	7.52
Phœbe	8,051,000	12,954,000	550.4	0.163	175

Diameter		Escape velocity		Magnitude	Max. apparent diameter as seen from planet
miles	km	mph	kph		
17 × 14 × 11	27 × 22 × 18	0.01	0.016	11.6	12′.3″
6 × 7 × 9	10 × 12 × 15	0.005	0.008	12.8	1′.7″
25	40	0.010?	0.02?	17.4	
16 × 12 × 10	26 × 20 × 16	0.006?	0.01?	18.9	
163 × 91 × 83	262 × 146 × 134	0.010?	0.16?	14.1	7′ 24″
68 × 62 × 56	110 × 100 × 90	0.006?	0.1?	15.5	
2,264	3,642	1.590	2.56	5.0	35′ 40″
1,945	3,130	1.310	2.10	5.3	17′ 30″
3,274	5,268	1.730	2.78	4.6	18′ 06″
2,987	4,806	1.510	2.43	5.6	9′ 30″
6	10	0.003?	0.005?	20.2	0″.2
106	170	0.060?	0.1?	14.8	8″.2
15	24	0.006?	0.01?	18.4	0″.1
50	80	0.030?	0.05?	16.7	1″.4
12	20	0.006?	0.01?	18.9	0″.02?
19	30	0.010?	0.02?	18.0	0″.02?
22	36	0.01?	0.02?	17.7	0″.02?
17	28	0.006?	0.01?	18.3	0″.01?
12	20				
23 × 21 × 17	37 × 34 × 27				
23 × 21 × 17	37 × 34 × 27			18.1	
92 × 62 × 42	148 × 100 × 68			16.5	
68 × 55 × 39	110 × 88 × 62			16.3	
86 × 68 × 68	138 × 110 × 110			15.5	
118 × 121 × 96	190 × 194 × 154			14.5	
247	398	0.06	0.1	12.9	10′ 54″
310	498	0.01	0.2	11.8	10′ 36″
650	1,046	0.25	0.4	10.3	17′ 36″
19 × 16 × 10	30 × 26 × 16			19.0	
19 × 10 × 10	30 × 16 × 16			18.5	
696	1,120	0.56	0.9	10.4	12′ 24″
22 × 21 × 17	36 × 34 × 28			18.5	
950	1,528	0.37	0.6	9.7	10′ 42″
3,201	5,150	1.54	2.47	8.4	17′ 10″
224 × 174 × 140	360 × 280 × 226	0.01	0.2	14.2	43″
892	1,436	0.44	0.7	10 (var)	1′ 48″
137	220	0.06	0.1	16.5	3″

Satellite	Mean distance from centre of planet		Orbital period (days)	Orbital eccentricity	Orbital inclination to planet's equator (degrees)
	miles	km			
Uranus					
Cordelia	30,746	49,471	0.330	very low	very low
Ophelia	33,434	53,796	0.372	very low	very low
Bianca	36,776	59,173	0.433	very low	very low
Cressida	38,395	61,777	0.463	very low	very low
Desdemona	38,953	62,676	0.475	very low	very low
Juliet	39,995	64,352	0.493	very low	very low
Portia	41,072	66,085	0.513	very low	very low
Rosalind	41,604	69,941	0.558	very low	very low
Belinda	46,773	75,258	0.622	very low	very low
Puck	53,450	86,000	0.762	very low	very low
Miranda	80,423	129,400	1.414	0.017	0.0
Ariel	118,710	191,000	2.520	0.003	0.0
Umbriel	165,500	266,300	4.144	0.004	0.0
Titania	270,300	435,000	8.706	0.002	0.0
Oberon	362,600	583,500	13.463	0.001	0.0
Neptune					
Naiad	29,960	48,200	0.30	very low	4.5
Thalassa	31,100	50,000	0.31	very low	very low
Despina	32,600	52,500	0.33	very low	very low
Galatea	38,500	62,000	0.40	very low	very low
Larissa	45,700	73,600	0.56	very low	very low
Proteus	73,089	117,600	1.12	very low	very low
Triton	220,000	355,300	5.877	0	159.9
Nereid	3,500,000	5,510,000	359.881	0.749	27.2
Pluto					
Charon	12,100	19,460	6.39	0	99

Diameter		Escape velocity		Magnitude	Max. apparent diameter as seen from planet
miles	km	mph	kph		
16	26	very low	very low		
19	30	very low	very low		
26	42	very low	very low		
39	62	very low	very low		
34	54	very low	very low		
52	84	very low	very low		
67	108	very low	very low		
34	54	very low	very low		
41	66	very low	very low		
96	154	very low	very low		
293	472	0.3?	0.5?	16.5	17' 54"
720	1,158	0.8	1.2	14.4	30' 54"
727	1,169	0.8	1.2	15.3	14' 12"
981	1,578	1.0	1.6	14.0	15' 00"
947	1,523	1.0	1.5	14.2	9' 48"
34	54				
50	80				
112	180				
93	150				
119	192				
258	416				
1,681	2,705	0.9	1.44	13.6	1° 01"
149	240			18.7	average 19"
790	1,270	0.1	0.16	17	

APPENDIX III
LUNAR DATA

Distance from Earth, centre to centre:

| km | max. 406,697 | min. 356,410 | mean 384,400 |
| miles | 252,764 | 221,510 | 238,906 |

Distance from Earth, surface to surface:

| km | max. 398,581 | min. 348,294 | mean 376,284 |
| miles | 247,720 | 216,466 | 233,862 |

Revolution period: 27.321 days

Axial rotation period: 27.321 days

Synodic period: 29 days 12 hr 44 min 3 sec

Mean velocity in orbit: 2,287 mph (3,680 kph)

Axial inclination of equator, relative to ecliptic: 1° 32′

Orbital inclination: 5° 09′

Diameter: 2,160.1 miles (3,475.6 km)

Apparent diameter as seen from Earth: max. 33′ 31″, min. 29′ 22″, mean 31′ 5″.

Mass (Earth = 1): 1/81.3

Density (water = 1): 3.34

Volume (Earth = 1): 0.02

Escape velocity: 1.48 miles/sec (2.38 km/sec)

Surface gravity (Earth = 1): 0.165

APPENDIX IV

LUNAR PROBES (1958–1994)

Probe	Launch date	Landing site (lat. long.)	
Russian			
Luna 1	2 Jan. 1959	–	Passed Moon at 4,660 miles (7,500 km). Data returned. In solar orbit.
Luna 2	12 Aug. 1959	?30N, ?1W	Crash-landing; exact site uncertain.
Luna 3	4 Oct. 1959	–	Obtained first photographs of the far side.
Luna 4	2 Apr. 1963	–	Failure; missed Moon by 5,280 miles (8,500 km). In solar orbit.
Cosmos 60	12 Mar. 1965	–	Probably an unsuccessful lunar probe.
Luna 5	9 May 1965	31S, 8E	Crashed in Mare Nubium. Attempted soft-lander.
Luna 6	8 June 1965	–	Missed Moon by 100,000 miles (161,000 km) on 11 June. In solar orbit.
Zond 3	18 July 1965	–	Passed Moon at 5,730 miles (9,219 km) and returned 25 pictures of the far side. In solar orbit.
Luna 7	4 Oct. 1965	9N, 40W	Crashed in Oceanus Procellarum. Unsuccessful soft-lander.
Luna 8	3 Dec. 1965	9.1N, 63.3W	Crashed in Oceanus Procellarum. Unsuccessful soft-lander.
Luna 9	31 Jan. 1966	7.1N, 64.4W	Successful soft-lander (Oceanus Procellarum). Photographs returned.
Luna 10	31 Mar. 1966	–	Lunar orbiter; minimum distance from Moon, 218 miles (350 km). Contact maintained for 460 orbits.
Luna 11	24 Aug. 1966	–	Lunar orbiter; minimum distance from Moon, 99 miles (159 km). Contact maintained until 1 Oct. 1966 (277 orbits).
Luna 12	22 Oct. 1966	–	Lunar orbiter; minimum distance from Moon, 62 miles (100 km). Transmitted photographs with resolution down to 50 feet (15 m). Contact maintained until Jan. 1967 (602 orbits).
Luna 13	21 Dec. 1966	18.9N, 62W	Successful soft-lander; Oceanus Procellarum. Pictures transmitted; soil analysis, surface radioactivity. Contact maintained until 27 Dec.
Zond 4	2 Mar. 1968	–	Probably a lunar probe; no precise information.
Luna 14	7 Apr. 1968	–	Lunar orbiter; minimum distance from Moon 100 miles (160 km). Gravitational and surface studies.
Zond 5	15 Sept. 1968	–	First probe to go round the Moon and return to Earth. Approached Moon to 1,210 miles (1,950 km). Seeds, insects, and even tortoises carried. Recovered from Indian Ocean on 21 Sept.
Zond 6	10 Nov. 1968	–	Circum-lunar probe; minimum distance from Moon 1,504 miles (2,420 km). General data, including far-side photographs. Landed safely in the USSR on 17 Nov.
Luna 15	13 July 1969	17N, 60E	Unsuccessful sample-and-return probe; crashed in Mare Crisium on 21 July, after 52 orbits of the Moon.
Zond 7	7 Aug. 1969	–	Lunar orbiter; minimum distance from Moon, 1,240 miles (2,000 km). Colour photographs of both Earth and Moon

Luna 16	12 Sept. 1970	7S, 55.3E	were brought back. Landed safely on Earth on 14 Feb. Successful sample-and-return probe. Landed in the Mare Fœcunditatis, and brought back 3 oz (100 g) of lunar material; a drill was used to collect material from a depth of 11 inches (30 cm). Landed safely back on Earth on 24 Sept.
Zond 8	20 Oct. 1970	–	Lunar orbiter; photographic data. Returned to Earth, splashing down safely in the Indian Ocean on 27 Oct.
Luna 17	10 Nov. 1970	38.3N, 35W	Landed in Mare Imbrium, carrying Lunokhod 1, which operated for 11 months, travelling 6.5 miles (10.5 km), and returning over 20,000 pictures as well as undertaking soil analysis and experiments of all kinds.
Luna 18	2 Sept. 1971	3.6N, 56.5E	Crashed in Mare Fœcunditatis after having made 54 orbits of the Moon. Possibly an intended sample-and-return probe.
Luna 19	28 Sept. 1971	–	Lunar orbiter; photographic and gravitational studies. Contact maintained until 3 Oct. 1972, after over 1,000 orbits of the Moon.
Luna 20	14 Feb. 1972	3.5N, 56.6E	Landed on the edge of the Mare Fœcunditatis, 75 miles (120 km) north of the landing point of Luna 16. Photographic studies. Samples obtained, including some from a drilled depth of 6 inches (150 mm). Luna 20 landed safely back on Earth on 25 Feb.
Luna 21	8 Jan. 1973	26N, 31E	Landed near Le Monnier crater, carrying Lunokhod 2; landing area 112 miles (180 km) from that of Apollo 17. Lunokhod 2 operated until 3 June, covering 23 miles (37 km) and sending back over 80,000 pictures. It set up a reflector which has been located by Earth-based laser equipment.
Luna 22	29 May 1974	–	Lunar orbiter; minimum distance from Moon, 15.5 miles (25 km). General studies. Contact maintained until Oct. 1975, after over 4,000 orbits of the Moon.
Luna 23	28 Oct. 1974	Uncertain	Soft-landed in Mare Crisium; unsuccessful sample-and-return probe. Contact lost on 9 Nov.
Luna 24	9 Aug. 1976	12.8N, 62.2E	Landed in Mare Crisium. Samples obtained, including those from a depth of 6.5 feet (2 m). Luna 24 landed safely back on Earth on 22 Aug.

American

Able 1	17 Aug. 1958	–	Failed to enter orbit.
Pioneer 1	11 Oct. 1958	–	Failed to reach the Moon; achieved 70,760 miles (113,854 km). Sent back data for 43 hr 17 min.
Pioneer 2	8 Nov. 1958	–	Failure; 3rd stage did not ignite.
Pioneer 3	6 Dec. 1958	–	Failed to reach the Moon; achieved 63,600 miles (102,333 km). Radiation data sent back.
Pioneer 4	3 Mar. 1959	–	Passed Moon at 37,280 miles (59,983 km). In solar orbit.
Able 4	26 Nov. 1959	–	Failure.
Able 5	25 Sept. 1960	–	Failure.
Able 5B	15 Dec. 1960	–	Failure.
Ranger 1	23 Aug. 1961	–	Failed to leave Earth orbit.
Ranger 2	18 Nov. 1961	–	Failed to leave Earth orbit.
Ranger 3	26 Jan. 1962	–	Missed Moon by 22,870 miles (36,800 km).
Ranger 4	23 Apr. 1962	15.5S, 130.7W	Crash-landed in Oceanus Procellarum. Camera failure.
Ranger 5	18 Oct. 1962	–	Missed Moon by 451 miles (725 km). In solar orbit.
Ranger 6	30 Jan. 1964	0.2N, 21.5E	Crash-landed in Mare Tranquillitatis. Camera failure.
Ranger 7	28 July 1964	10.7S, 20.7W	Crash-landed in Mare Nubium. Successful photographic probe; 4,308 pictures returned.

Ranger 8	17 Feb. 1965	2.7N, 24.8E	Crash-landed in Mare Tranquillitatis. Successful photographic probe; 7,137 pictures returned.
Ranger 9	21 Mar. 1965	12.9S, 2.4W	Crash-landed in Alphonsus. Successful photographic probe; 5,814 pictures returned.
Surveyor 1	30 May 1966	2.5S, 43.2W	Landed near Flamsteed; returned 11,237 pictures.
Orbiter 1	10 Aug. 1966	6.7N, 162E	Successful photographic probe. Finally crash-landed on 29 Oct. 1966.
Surveyor 2	20 Sept. 1966	5N, 25W	Unsuccessful soft-lander; crashed near Copernicus.
Orbiter 2	6 Nov. 1966	4S, 98E	Successful photographic probe. 422 pictures returned.
Orbiter 3	4 Feb. 1967	14.6N, 91.7W	Successful photographic probe. 307 pictures returned.
Surveyor 3	17 Apr. 1967	2.9S, 23.3W	Landed in Oceanus Procellarum. 6,315 pictures returned. Soil analyses.
Orbiter 4	4 May 1967	Far side	Successful photographic probe. 326 pictures returned.
Surveyor 4	14 July 1967	0.4N, 1.3W	Unsuccessful soft-lander. Crashed in Sinus Medii.
Explorer 35	19 July 1967	–	Studies of magnetic fields near Moon and Earth.
Orbiter 5	1 Aug. 1967	0, 70W	Successful photographic probe. Controlled impact on 31 Jan. 1968.
Surveyor 5	8 Sept. 1967	1.4N, 23.2E	Landed in Mare Tranquillitatis. 18,006 pictures returned.
Surveyor 6	7 Nov. 1967	0.5N, 1.4W	Landed in Sinus Medii. 30,065 pictures returned.
Surveyor 7	7 Jan. 1968	40.9S, 11.5W	Landed on north rim of Tycho. 21,274 pictures returned. Soil analyses.
Apollo 8	21 Dec. 1968	–	Manned orbiter (F. Borman, J. Lovell, W. Anders.) 10 orbits of Moon.
Apollo 10	18 May 1969	–	Manned orbiter (T. Stafford, E. Cernan, J. Young). 31 orbits of Moon.
Apollo 11	16 July 1969	00° 67'N, 23° 49'E	Landed in Mare Tranquillitatis, 20 July (N. Armstrong, E. Aldrin; M. Collins in orbit). 44 lbs (20 kg) samples returned.
Apollo 12	14 Nov. 1969	03° 12'S, 23° 23'W	Landed in Oceanus Procellarum, 19 Nov. (C. Conrad, M. Bean; R. Gordon in orbit). 75 lbs (34 kg) samples returned; Surveyor 3 visited.
Apollo 13	11 Apr. 1970	–	Unsuccessful manned lander (J. Lovell, F. Haise, J. Swigert). Went round Moon, and splashed down on Earth on 17 Apr.
Apollo 14	31 Jan. 1971	03° 40'S, 17° 28'W	Landed near Fra Mauro, 5 Feb. (A. Shepard, E. Mitchell; S. Roosa in orbit). Moon cart carried. 97 lbs (44 kg) samples returned.
Apollo 15	26 July 1971	26° 06'N, 03° 39'E	Landed in Hadley-Apennines, 30 July (D. Scott, J. Irwin; A. Worden in orbit). 172 lbs (78 kg) samples returned. Lunar Rover carried to the Moon.
Apollo 16	16 Apr. 1972	08° 36'S, 15° 31'E	Landed in Descartes area, 21 April (J. Young, C. Duke; C. Mattingly in orbit). 215 lbs (97.5 kg) samples returned. Lunar Rover carried to the Moon.
Apollo 17	6 Dec. 1972	20° 10'N, 30° 46'E.	Landed in Taurus–Littrow, 11 Dec. (E. Cernan, H. Schmitt; D. Evans in orbit). 249 lbs (113 kg) samples returned. Lunar Rover carried to the Moon. Geological studies.
Explorer 49	10 June 1973	–	Lunar orbiter; closest approach to Moon, 689 miles (1,109 km). Studies of radio conditions over the Moon's far side.
Explorer 50	26 Oct. 1973	–	Lunar orbiter; closest approach to Moon, 58,855 miles (94,697 km). Studies of magnetic fields.
Clementine	24 Jan. 1994	–	Lunar mapping probe.

Japanese

Hagomoro	24 Jan. 1990	–	Lunar orbiter, separated from Muses–A rocket; orbit almost circular, 10,000 miles (16,090 km) above the Moon. General observations. Carried Hiten lander, which crash-landed April 1993 at 34°S, 55°E, near Furnerius.
Hiten			

APPENDIX V
INTERPLANETARY PROBES (1961–1993)

Probe	Launch date	Encounter date	* = USA ** = USSR
Mercury			
Mariner 10*	3 Nov. 1973	29 Mar. 1974	Passed Mercury at 168 miles (271 km).
		21 Sept. 1974	Passed Mercury at 29,800 miles (48,000 km).
		24 Mar. 1975	Passed Mercury at 198 miles (319 km).
Venus			
Venera 1**	12 Feb. 1961	19 May ?1961	Failure; contact lost, at distance of 4,661,300 miles (7,500,000 km) from Earth.
Mariner 1*	22 July 1962	–	Failed to enter orbit.
Mariner 2*	26 Aug. 1962	14 Dec. 1962	Successful fly-by. Passed Venus at 21,750 miles (35,000 km).
Zond 1**	2 Apr. 1964	?	Failure; contact lost.
Venera 2**	12 Nov. 1965	27 Feb. 1966	No data received. In solar orbit.
Venera 3**	16 Nov. 1965	1 Mar. 1966	Landed, but no data received.
Venera 4**	12 June 1967	18 Oct. 1967	Landed. Data transmitted during descent.
Mariner 5*	14 June 1967	19 Oct. 1967	Successful photographic fly-by.
Venera 5**	5 Jan. 1969	16 May 1969	Landed on the dark side of Venus. Data transmitted during descent.
Venera 6**	10 Jan. 1969	17 May 1969	Landed on the dark side of Venus. Data transmitted during descent.
Venera 7**	17 Aug. 1970	15 Dec. 1970	Landed; transmitted for 23 minutes after arrival.
Venera 8**	26 Mar. 1972	22 July 1972	Landed (10°S, long. 335°) 1,800 miles (2,900 km) from Venera 7 site. Transmitted for 50 minutes after arrival.
Mariner 10*	3 Nov. 1973	5 Feb. 1974	Passed Venus at 3,600 miles (5,800 km) *en route* to Mercury. Photographs received.
Venera 9**	8 June 1975	21 Oct. 1975	Landed at 31° 41′N, long. 293° 50′. One picture received. Transmitted for 53 minutes after arrival.
Venera 10**	14 June 1975	25 Oct. 1975	Landed at 16°N, long. 291°. One picture received. Transmitted for 65 minutes after arrival.
Pioneer Venus 1*	20 May 1978	4 Dec. 1978	Successful orbiter; radar mapping of Venus's surface.
Pioneer Venus 2*	8 Aug. 1978	9 Dec. 1978	Multi-probe; one bus and 4 entry probes. Bus burned away in the atmosphere at 33°S, long. 43°W. Sounder Probe landed at 0°, 43°W, and transmitted for 68 minutes after arrival. North Probe landed at 75°N, 20°E; Day Probe at 26°S, 45°W; and Night Probe at 27°S, 45°E.
Venera 11**	9 Sept. 1978	25 Dec. 1978.	Landed. Transmitted for about 1 hour. No pictures sent back. Landing site 14°S, long. 299°.
Venera 12**	14 Sept. 1978	21 Dec. 1978	Landed. Transmitted for about 1 hour. No pictures sent back. Landing site 7°S, long. 294°, about 500 miles (800 km) from the Venera 11 site.
Venera 13**	30 Oct. 1981	1 Mar. 1982	Landed. Transmitted for 127 minutes; 8 colour pictures received. Soil analyses. Landing site 7° 30′S, long. 303°, in Phœbe Regio.

Venera 14**	4 Nov. 1981	5 Mar. 1982	Landed. Transmitted for 57 minutes. Colour pictures, soil analyses. Landing site 13° 13′S, long. 310°, 620 miles (1,000 km) from Venera 13 site.
Venera 15**	2 June 1983	10 Oct. 1983	Successful polar orbiter. Radar mapping of surface.
Venera 16**	7 June 1983	16 Oct. 1983	Successful polar orbiter. Radar mapping of surface.
Vega 1**	15 Dec. 1984	11 June 1985	Passed Venus at 5,530 miles (8,890 km) *en route* to Halley's Comet. Balloon dropped into Venus' atmosphere.
Vega 2**	20 Dec. 1984	15 June 1985	Passed Venus at 4,990 miles (8,030 km) *en route* to Halley's Comet. Balloon dropped into Venus's atmosphere.
Magellan*	5 May 1989	10 Aug. 1990	Radar mapper, launched from Atlantis Shuttle.
Galileo	18 Oct. 1989	10 Feb. 1990	Flew past Venus at 9,940 miles (16,000 km) *en route* to Jupiter.

Mars

Mars 1**	1 Nov. 1962	?	Failure; contact lost. Probably approached Mars to 118,000 miles (190,000 km). In solar orbit.
Mariner 3*	5 Nov. 1964	–	Failure. Contact lost soon after launch. In solar orbit.
Mariner 4*	28 Nov. 1964	14 July 1965	Successful fly-by. Closest approach to Mars 6,200 miles (10,000 km). 21 pictures returned.
Zond 2**	30 Nov. 1964	Aug? 1965	Failure. Contact lost on 2 May 1965.
Mariner 6*	24 Feb. 1969	31 July 1969	Successful fly-by; passed Mars at 2,110 miles (3,390 km). 76 pictures returned.
Mariner 7*	27 Mar. 1969	4 Aug. 1969	Successful fly-by; passed Mars at 2,175 miles (3,500 km). 126 pictures returned.
Mariner 8*	8 May 1971	–	Failed to enter orbit.
Mariner 9*	30 May 1971	13 Nov. 1971	Successful orbiter. 7,329 pictures returned. Contact maintained until 27 Oct. 1972.
Mars 2**	19 May 1971	27 Nov. 1971	Orbiter and lander. Landing site 44°.2S, long. 213°.2, in Eridania, but contact lost.
Mars 3**	28 May 1971	2 Dec. 1971	Orbiter and lander. Landing site 45°S, long. 158°, in Phæthontis, but contact lost after 20 seconds; no data received.
Mariner 9*	30 May 1971	13 Nov. 1971	Successful orbiter. 7,329 pictures sent back. Contact maintained until 27 Oct. 1972.
Mars 4**	21 July 1973	10 Feb. 1974	Missed Mars by 1,370 miles (2,200 km), but some pictures sent back.
Mars 5**	25 July 1973	12 Feb. 1974	Orbiter. Contact soon lost, but some pictures sent back.
Mars 6**	5 Aug. 1973	12 Mar. 1974	Orbiter and lander. Landing site about 24°S, long. 25°, in Mare Erythræum; contact lost during landing sequence, but some data received.
Mars 7**	9 Aug. 1973	9 Mar. 1974	Missed Mars by 810 miles (1,300 km). In solar orbit.
Viking 1*	20 Aug. 1975	19 June 1976 (in orbit)	Successful lander and orbiter. Landing site 22°.4N, long. 47°.5, in Chryse. Contact with orbiter maintained until 7 Aug. 1980, and with lander until May 1983.
Viking 2*	9 Sept. 1975	7 Aug. 1976 (in orbit)	Successful lander and orbiter. Landing site 47°.9N, long. 225°.9, in Utopia. Contact with orbiter maintained until 24 July 1980, and with lander until 11 April 1980.
Phobos 1**	7 July 1988	?	Contact lost, 29 Aug. 1988. No data received.
Phobos 2**	12 July 1988	March 1989	Contact lost, 27 March 1989. Some pictures of Phobos received, and also some data.
Mars Observer*	25 Sept. 1992	24 Aug. 1993	Failure; contact lost on 25 Aug. 1993.

Jupiter

Pioneer 10*	2 Mar. 1972	3 Dec. 1973	Successful fly-by. Passed Jupiter at 81,670 miles (131,400 km).
Pioneer 11*	5 Apr. 1973	2 Dec. 1974	Successful fly-by. Passed Jupiter at 28,840 miles (46,400 km). Went on to fly by Saturn.
Voyager 1*	5 Sept. 1977	5 Mar. 1979	Successful fly-by. Passed Jupiter at 217,500 miles (350,000 km). Went on to fly by Saturn.
Voyager 2*	20 Aug. 1977	9 July 1979	Successful fly-by. Passed Jupiter at 443,700 miles (714,000 km). Went on to Saturn, Uranus and Neptune.
Galileo	18 Oct. 1989	Dec. 1995	Entry probe and orbiter.
Ulysses	6 Oct. 1990	8 Feb. 1992	Solar polar probe; passed Jupiter at 235,000 miles (378,000 km); studies of Jovian magnetosphere, radiation zones and general environment.

Saturn

Pioneer 11*	5 Apr. 1973	1 Sept. 1979	Successful fly-by. Passed Saturn at 12,980 miles (20,880 km).
Voyager 1*	5 Sept. 1977	12 Nov. 1980	Successful fly-by. Passed Saturn at 77,200 miles (124,200 km).
Voyager 2*	20 Aug. 1977	25 Aug. 1981	Successful fly-by. Passed Saturn at 62,960 miles (101,300 km). Went on to Uranus and Neptune.

Uranus

Voyager 2*	20 Aug. 1977	24 Jan. 1986	Successful fly-by. Passed Uranus at 49,700 (80,000 km). Went on to Neptune.

Neptune

Voyager 2*	20 Aug. 1977	26 Aug. 1989	Successful fly-by. Passed Neptune at 3,000 miles (5,000 km).

Comet Giacobini–Zinner

ICE*	12 Aug. 1978	11 Sept. 1985	Originally ISEE-3. Went through the comet's tail.

Halley's Comet

ICE*	12 Aug. 1978	25 Mar. 1986	Passed Halley's Comet at 18,600,000 miles (30,000,000 km), but no useful data obtained.
Vega 1**	15 Dec. 1984	6 Mar. 1986	Passed comet at 5,525 miles (8,890 km).
Vega 2**	21 Dec. 1984	9 Mar. 1986	Passed comet at 4,990 miles (8,030 km).
Sakigake	8 Jan. 1985	11 Mar. 1986	Japanese; passed comet at 93,850 miles (151,000 km).
Giotto	2 July 1985	14 Mar. 1986	European. Passed nucleus of comet at 370 miles (596 km).
Suisei	18 Aug. 1985	8 Mar. 1986	Japanese. Passed comet at 4,300,000 miles (7,000,000 km).

Comet Grigg-Skjellerup

Giotto	2 July 1985	10 July 1992	Passed comet's nucleus at about 200 miles (320 km).

INDEX